Apache Airflow Best Practices

A practical guide to orchestrating data workflow
with Apache Airflow

Dylan Intorf

Dylan Storey

Kendrick van Doorn

Apache Airflow Best Practices

Copyright © 2024 Packt Publishing

The authors acknowledge the use of cutting-edge AI, such as ChatGPT, with the sole aim of enhancing the language and clarity within the book, thereby ensuring a smooth reading experience for readers. It's important to note that the content itself has been crafted by the authors and edited by a professional publishing team.

Every effort has been made in the preparation of this book to ensure the accuracy of the information presented. However, the information contained in this book is sold without warranty, either express or implied. Neither the authors, nor Packt Publishing or its dealers and distributors, will be held liable for any damages caused or alleged to have been caused directly or indirectly by this book.

Packt Publishing has endeavored to provide trademark information about all of the companies and products mentioned in this book by the appropriate use of capitals. However, Packt Publishing cannot guarantee the accuracy of this information.

Group Product Manager: Ali Abidi

Publishing Product Manager: Apeksha Shetty

Book Project Manager: Shambhavi Mishra

Senior Editor: Joseph Sunil

Technical Editor: Seemanjay Ameriya

Copy Editor: Safis Editing

Proofreader: Joseph Sunil

Indexer: Pratik Shirodkar

Production Designer: Joshua Misquitta

Senior DevRel Marketing Executive: Vinishka Kalra

First published: October 2024

Production reference: 1101024

Published by Packt Publishing Ltd.

Grosvenor House

11 St Paul's Square

Birmingham

B3 1RB, UK.

ISBN 978-1-80512-375-0

www.packtpub.com

To my partner, Kristen, for always supporting my dreams and encouraging me to have confidence the pipelines won't break in the middle of the night.

– Kendrick van Doorn

Contributors

About the authors

Dylan Intorf is a seasoned technology leader with a B.Sc. in computer science from Arizona State University. With over a decade of experience in software and data engineering, he has delivered custom, tailored solutions to the technology, financial, and insurance sectors. Dylan's expertise in data and infrastructure management has been instrumental in optimizing Airflow deployments and operations for several Fortune 25 companies.

Dylan Storey holds a B.Sc. and M.Sc. in biology from California State University, Fresno, and a Ph.D. in life sciences from the University of Tennessee, Knoxville where he specialized in leveraging computational methods to study complex biological systems. With over 15 years of experience, Dylan has successfully built, grown, and led teams to drive the development and operation of data products across various scales and industries, including many of the top Fortune-recognized organizations. He is also an expert in leveraging AI and machine learning to automate processes and decisions, enabling businesses to achieve their strategic goals.

Kendrick van Doorn is an accomplished engineering and business leader with a strong foundation in software development, honed through impactful work with federal agencies and consulting technology firms. With over a decade of experience in crafting technology and data strategies for leading brands, he has consistently driven innovation and efficiency. Kendrick holds a B.Sc. in computer engineering from Villanova University, an M.Sc. in systems engineering from George Mason University, and an MBA from Columbia University.

About the reviewers

Ayoade Adegbite is an accomplished data and analytics engineer with extensive experience in leveraging advanced data tools and enterprise ecosystems to deliver actionable insights across diverse industries. He excels in designing sophisticated analytical models, ensuring data integrity, and implementing impactful data solutions. With a strong background in ETL processes, data visualization, and robust documentation practices, Ayoade has consistently driven significant improvements in data-driven decision-making and operational efficiency.

Ayoade has also optimized business processes and revitalized operations as a consultant, utilizing technologies such as Airflow and dbt. Ayoade is a member of the Apache Airflow Champion initiative.

Ananth Packkildurai, the author of the influential Data Engineering Weekly newsletter, has made significant contributions to the data industry through his deep expertise and innovative insights. His work has been instrumental in shaping how companies approach data engineering, and industry leaders highly regard his thought leadership. In the past, Ananth worked in companies such as Slack and Zendesk to build petabyte-scale data infrastructure, including a data pipeline, search infrastructure, customer-facing analytics, and an observability platform.

Frank Breetz is a highly experienced data consultant with over a decade of expertise in the field. His proficiency in Apache Airflow is extensive and well-rounded. During his tenure at Astronomer, Frank advised numerous clients on Airflow best practices and helped establish industry standards. Currently, at LinQuest Corporation, he continues to leverage Airflow alongside various other technologies while developing a Data Management Framework.

Frank's in-depth knowledge and practical experience make him an invaluable resource for mastering Apache Airflow. He holds a Master's degree in Computer Science and a Bachelor's degree in Physics.

Vipul Bharat Marlecha is an accomplished software engineer with a focus on large-scale distributed data systems. His career spans roles at major tech companies, including Netflix, DoorDash, Twitter, and Nomura. He is particularly skilled in managing big data, designing scalable systems, and delivering solutions that emphasize impact over activity.

Table of Contents

Part 2: Airflow Basics

3

Components of Airflow   29

4

Basics of Airflow and DAG Authoring   43

Part 3: Common Use Cases

5

6

Part 4: Scale with Your Deployed Instance

10

11

Preface

This book serves as a practical guide to mastering Apache Airflow in production environments, offering insights into its deployment, management, and optimization. As data pipelines become increasingly complex, ensuring reliability, scalability, and efficiency in orchestrating tasks is crucial for any data-driven organization. Through hands-on examples and best practices, this book equips you with the knowledge to navigate the challenges of operating Airflow at scale, from initial setup to continuous monitoring and improvement.

Who this book is for

This book is designed for engineering leads, platform engineers, data engineers, and technical product/platform managers who are responsible for building, maintaining, and scaling data infrastructure in production environments. It caters to those who already have a foundational understanding of data pipelines and are looking to deepen their expertise in Apache Airflow, particularly in the context of large-scale, real-world applications. Whether you are overseeing a team, managing a data platform, or directly implementing workflows, this book provides the strategic insights and practical tools needed to ensure your Airflow deployments are robust, efficient, and aligned with your organizational goals.

What this book covers

Chapter 1, Getting Started with Airflow 2.0, offers a brief introduction to what Airflow is and why it is an important component of today's data ecosystem.

Chapter 2, Core Airflow Concepts, covers core concepts, objects, and ontologies important for understanding and describing how Airflow functions.

Chapter 3, Components of Airflow, covers core operating components of Airflow, what they are responsible for, and generally how they work.

Chapter 4, Basics of Airflow and DAG Authoring, covers the basic process of writing a workflow in Airflow.

Chapter 5, Connecting to External Sources, covers how to set up and configure external sources of information/interaction with Airflow.

Chapter 6, Extending Functionality with UI Plugins, covers how to customize Airflow's UI with a custom UI plugin, allowing for streamlined organization-specific operational needs.

Chapter 7, Writing and Distributing Custom Providers, covers the process of how to apply software engineering principles to write, test, and publish Airflow providers.

Chapter 8, Orchestrating a Machine Learning Workflow, demonstrates a relatively common use case in Airflow of orchestrating the training and delivery of a machine learning model.

Chapter 9, Using Airflow as a Driving Service, demonstrates how Airflow can be used as a driving service, allowing consumers to use an abstracted interface instead of directly writing workflows.

Chapter 10, Airflow Ops: Development and Deployment, covers and demonstrates how to customize your Airflow deployment, apply version control to your workflows, and develop a development life cycle for the delivery and maintenance of workflows.

Chapter 11, Airflow Ops Best Practices: Observation and Monitoring, covers practices for monitoring and maintaining your core Airflow deployment and your operational DAGs.

Chapter 12, Multi-Tenancy in Airflow, covers how to design, configure, and operate your Airflow deployment to achieve differing levels of multi-tenancy and compute isolation.

Chapter 13, Migrating Airflow, discusses practical guidance in planning and executing a migration of your Airflow instance from one environment to another.

To get the most out of this book

The code sources and examples in this book were primarily developed with the assumption that you would have access to Docker and Docker Compose. We also make some assumptions that you have a passing familiarity with Python, Kubernetes, and Docker.

Software/hardware covered in the book	Operating system requirements
Airflow 2.0+	Windows, macOS, or Linux
Python 3.9+	
Docker	
Postgres	

If you are using the digital version of this book, we advise you to type the code yourself or access the code from the book's GitHub repository (a link is available in the next section). Doing so will help you avoid any potential errors related to the copying and pasting of code.

Download the example code files

You can download the example code files for this book from GitHub at https://github.com/PacktPublishing/Apache-Airflow-Best-Practices. If there's an update to the code, it will be updated in the GitHub repository.

We also have other code bundles from our rich catalog of books and videos available at https://github.com/PacktPublishing/. Check them out!

Conventions used

There are a number of text conventions used throughout this book.

`Code in text`: Indicates code words in text, database table names, folder names, filenames, file extensions, pathnames, dummy URLs, user input, and Twitter handles. Here is an example: "Once you have checked that all services are up and running by navigating to `localhost:8080`, you can log in to your Airflow instance and see (and even operate) your DAG."

A block of code is set as follows:

```
class MetricsPlugin(AirflowPlugin):
    """Defining the plugin class"""
    name = "Metrics Dashboard Plugin"
    flask_blueprints = [metrics_blueprint]
    appbuilder_views = [{
        "name": "Dashboard", "category": "Metrics",
        "view": MetricsDashboardView()
    }]
```

Any command-line input or output is written as follows:

```
$ pip install airflowctl
```

Bold: Indicates a new term, an important word, or words that you see onscreen. For instance, words in menus or dialog boxes appear in **bold**. Here is an example: "To find the connections section of Apache Airflow, navigate to the top bar of the Airflow UI. By selecting **Admin**, the drop-down menu will show **Connections**."

> **Tips or important notes**
> Appear like this.

Get in touch

Feedback from our readers is always welcome.

General feedback: If you have questions about any aspect of this book, email us at `customercare@packtpub.com` and mention the book title in the subject of your message.

Errata: Although we have taken every care to ensure the accuracy of our content, mistakes do happen. If you have found a mistake in this book, we would be grateful if you would report this to us. Please visit www.packtpub.com/support/errata and fill in the form.

Piracy: If you come across any illegal copies of our works in any form on the internet, we would be grateful if you would provide us with the location address or website name. Please contact us at copyright@packt.com with a link to the material.

If you are interested in becoming an author: If there is a topic that you have expertise in and you are interested in either writing or contributing to a book, please visit authors.packtpub.com.

Share your thoughts

Once you've read *Apache Airflow Best Practices*, we'd love to hear your thoughts! Scan the QR code below to go straight to the Amazon review page for this book and share your feedback.

https://packt.link/r/1-805-12375-0

Your review is important to us and the tech community and will help us make sure we're delivering excellent quality content.

Download a free PDF copy of this book

Thanks for purchasing this book!

Do you like to read on the go but are unable to carry your print books everywhere?

Is your eBook purchase not compatible with the device of your choice?

Don't worry, now with every Packt book you get a DRM-free PDF version of that book at no cost.

Read anywhere, any place, on any device. Search, copy, and paste code from your favorite technical books directly into your application.

The perks don't stop there, you can get exclusive access to discounts, newsletters, and great free content in your inbox daily

Follow these simple steps to get the benefits:

1. Scan the QR code or visit the link below

https://packt.link/free-ebook/978-1-80512-375-0

2. Submit your proof of purchase
3. That's it! We'll send your free PDF and other benefits to your email directly

Part 1:
Apache Airflow:
History, What, and Why

This part has the following chapters:

1

Getting Started with Airflow 2.0

In modern software development and data processing, orchestration plays a pivotal role in ensuring the coordination and execution of complex workflows. As organizations strive to manage their ever-growing data and application landscapes, the need for an efficient orchestration system becomes paramount.

With Airflow 2.0 having been released for some time and moving quickly to increase its capabilities, we elected to distill our experiences in operating Airflow to help others by showing them patterns that have worked well for others in the past.

Our goal with this book is to help engineers and organizations adopting Apache Airflow as their orchestration solution get the most out of their technology selection by guiding them to better choices as they go through their adoption journey and scale.

In this chapter, we will learn what data orchestration is and how it is applied to several industries facing data challenges. In addition, we will explore the basic benefits of Apache Airflow and its features that may benefit your organization. We will take a look ahead at what you can expect to learn by reading this book and practicing industry-leading techniques for orchestrating your data pipelines with Apache Airflow. Apache Airflow remains the industry leader in data orchestration and pipeline management. With this success comes a set of tenets and principles that have been identified as best practices. We will cover some of the best practices and approaches in this chapter and identify the skills needed to be successful.

In this chapter, we're going to cover the following main topics:

- What is data orchestration?
- Exploring Apache Airflow
- Core concepts of Airflow
- Skills to use Apache Airflow effectively

What is data orchestration?

In today's data-driven world, organizations face the challenge of handling vast amounts of data from diverse sources. Data orchestration is the key to managing this complex data landscape efficiently. It involves the coordination, automation, and monitoring of data workflows, ensuring the smooth execution of tasks and the timely delivery of valuable insights.

Orchestration, in the context of software development and data engineering, refers to the process of automating and managing the execution of interconnected tasks or processes to achieve a specific goal. These tasks might involve data processing, workflow scheduling, service provisioning, and more. The purpose of orchestration is to streamline the flow of operations, optimize resource utilization, and ensure that tasks are executed in a well-coordinated manner.

Traditional, manual orchestration is cumbersome and prone to errors, especially as the complexity of workflows increases. However, with modern orchestration tools and frameworks, developers can automate these intricate processes, resulting in enhanced efficiency and reliability.

Industry use cases

Regardless of the industry, Apache Airflow can bring benefits to any data engineering or data analysis team. To better align this, here are some examples of how a few key industries that we have worked with in the past may use this leading data orchestrator to benefit their needs:

- **E-commerce**: An e-commerce brand may need an automated ETL/ELT pipeline for automating the extraction, transformation, and loading of data from various sources, such as sales, customer interactions, and current inventory
- **Banking/fintech**: Leading financial firms may use Apache Airflow to orchestrate the processing of transaction data to identify fraud or risks in their reporting/billing systems
- **Retail**: Major retailers and brands can use Apache Airflow to help automate their **machine learning** (ML) workloads to better predict user trends and purchases based on seasonality or current market environments

Now that we have learned what data orchestration is, how it is important for organizations, and some basic industry use-case examples, let us explore Apache Airflow, which is one of the most popular platforms and the core topic of this book.

Exploring Apache Airflow

Apache Airflow is known within the data engineering community as the go-to open source platform for "*developing, scheduling, and monitoring batch-oriented workflows.*" (Apache.org Airflow documentation: `https://airflow.apache.org/docs/apache-airflow/stable/index.html`)

Apache Airflow has emerged as the go-to open source platform for data orchestration and remains the leader as a result of its active development community. It offers a robust and flexible solution to the challenges of managing complex data workflows. Airflow enables data engineers, data scientists, **artificial intelligence** (**AI**)/ML engineering, and MLOps and DevOps professionals to design, schedule, and monitor data pipelines with ease.

The power of Apache Airflow lies in its ability to represent data workflows as **directed acyclic graphs** (**DAGs**). This intuitive approach allows users to visualize and understand relationships between tasks, making it easier to create and maintain complex data pipelines. Furthermore, Airflow's extensibility and modularity allow users to customize the platform to their specific needs, making it an ideal choice for businesses of all sizes and industries.

Apache Airflow 2.0

The release of Apache Airflow 2 in December 2020 stands as one of the largest achievements of the community since Airflow was originally created as a solution at Airbnb in 2014. The move to 2.0 was a large lift for the community and came with hundreds of updates and bug fixes after an Airflow community survey in 2019.

This release brought with it an updated UI, a new scheduler, a Kubernetes executor, and a simpler way to group tasks of DAGs together. It was a groundbreaking achievement and laid out the roadmap for future releases that have only made Airflow an even more valuable tool for the community.

Standout features of Apache Airflow

Apache Airflow has brought with it a multitude of features to support the different needs of organizations and teams. Some of our favorites revolve around sensing, task grouping, and operators, but each of these can be grouped into one of these categories:

- **Extensible**: Users can create custom operators and sensors or access a wide range of community-contributed plugins, enabling seamless integration with various technologies and services. This extensibility enhances Airflow's adaptability to different environments and use cases, making its potential limited only by the engineer's imagination.

- **Dynamic**: The platform supports dynamic workflows, meaning the number of tasks and their configurations can be determined at runtime, based on variables, external sensors, or data captured during a run. This feature makes Airflow more flexible as workflows can adapt to changing conditions or input parameters, resulting in better resource utilization and improved efficiency.

- **Scalable**: Airflow's distributed architecture ensures scalability to handle large-scale and computationally intensive workflows. As businesses grow and their data processing demands increase, Airflow can accommodate these requirements by distributing tasks across multiple workers, reducing processing times, and improving overall performance.

- **Built-in monitoring**: Airflow provides a web-based UI to monitor the status of workflows and individual tasks. This interface allows users to visualize task execution and inspect logs, facilitating transparency and easy debugging. By gaining insights into workflow performance, users can optimize their processes and identify potential bottlenecks.

- **Ecosystem**: Airflow seamlessly integrates with a wide range of technologies and cloud providers. This integration allows users to access diverse data sources and services, making it easier to design comprehensive workflows that interact with various systems. Whether working with databases, cloud storage, or other tools, Airflow can bridge the gap between different components.

Apache Airflow brings with it years of open source development and well-thought-out designs by hundreds of contributors. It is the leading data orchestration tool, and learning how to better utilize its key features will help you become a better data engineer and manager.

A look ahead

Throughout this book, we will explore the essential features of Apache Airflow, providing you with the knowledge to leverage its full potential in your data orchestration journey. The key topics covered include the following:

- **Why use Airflow?**: Tenants, skills, and first principles

- **Airflow basics**: Understanding core concepts (DAGs, tasks, operators, deferrables, connections, and so on), the components of Airflow, and the basics of DAG authoring

- **Common use cases**: Unlocking the potential of Airflow with ETL pipelines, custom plugins, and orchestrating workloads across systems

- **Scaling with your team**: Hardening your Airflow instance for production workloads with CI/CD, monitoring, and the cloud

By the end of this book, you will have a comprehensive understanding of Apache Airflow's best practices, enabling you to build robust, scalable, and efficient data pipelines that drive your organization's success. Let's embark on this practical guide for data pipeline orchestration using Apache Airflow and unlock the true potential of data-driven decision-making.

Core concepts of Airflow

Apache Airflow is a dynamic, extensible, and flexible framework that allows for the building of workflows as code. Airflow allows for the definition of these automated workflows as code. This allows for better code versioning, development through CI/CD, easy testing, and extensible components and operators from a thriving community of committers.

Airflow is known for its approach to scheduling tasks and workflows. It can take advantage of CRON scheduling or its built-in scheduling functions. In addition, features such as backfilling, allowing for the rerun of pipelines, allow for going back and updating pipelines if the logic changes. This means it has powerful operational components that need to be accounted for as part of the design.

Following these guidelines will help you lay a foundation for scaling your Airflow deployments and increase the effectiveness of your workflows from both authorship and operational viewpoints.

Why Airflow may not be right

Before we jump into the tenets of Apache Airflow, we should take a moment to acknowledge when it is not a good fit for organizations. While each of the following statements can probably be shown to be "false" for sufficiently motivated and clever engineers (and we all know plenty of them), they are generally considered anti-patterns and should be avoided.

Some of these anti-patterns include the following:

- Teams where there is no or limited experience with Python programming. Implementing DAGs in Python can be a complex process and requires active experience to upkeep the code.

- Streaming or non-batch workflows and pipelines, where the use case requires immediate updates. Airflow is designed for batch-oriented and scheduled tasks.

When to choose Airflow

Airflow is best used for implementing "batch-oriented" scheduled data pipelines. Often, use cases can include ETL/ELT, reverse ETL, ML, AI, and **business intelligence** (**BI**). Throughout this book, we will review major use cases that have been seen at different industry-leading companies. Some key use cases include the following:

- **ETL pipelines of data**: Almost every implementation of Airflow helps to automate tasks of this type, whether it is to consolidate data within a data warehouse or move data through different tools

- **Writing and distributing custom plugins for organizations that have a unique stack and needs that have not been addressed by the open source community**: Airflow allows for easy customization of the environment and ecosystem to your needs

- **Extending the UI functionality with plugins**: Modify and adjust the UI to allow for new views, charts, and widgets to integrate with external systems

- **ML workflows and orchestrating across systems**: Teams building and maintaining ML workflows often depend on Airflow to complete the automation of training, transforming, and evaluating models

Each of these use cases requires a different set of basic skills to be implemented and scaled in support of a large organization.

Zen of Python

The *Zen of Python* is a great guide for any Python programmer to make better decisions when writing code; some portions of it are particularly good for data engineers to consider when working with Airflow. It is a collection of 19 "guiding principles" for writing Python from software engineering pioneer Tim Peters. Before going into the details on what we think is most valuable for Apache Airflow development, let's review the *Zen of Python*:

"Beautiful is better than ugly.

Explicit is better than implicit.

Simple is better than complex.

Complex is better than complicated.

Flat is better than nested.

Sparse is better than dense.

Readability counts.

Special cases aren't special enough to break the rules.

Although practicality beats purity.

Errors should never pass silently.

Unless explicitly silenced.

In the face of ambiguity, refuse the temptation to guess.

There should be one-- and preferably only one --obvious way to do it.

Although that way may not be obvious at first unless you're Dutch.

Now is better than never.

Although never is often better than *right* now.

If the implementation is hard to explain, it's a bad idea.

If the implementation is easy to explain, it may be a good idea.

Namespaces are one honking great idea -- let's do more of those!"

A few key guiding principles that should help in your learning are the following:

- *Explicit is better than implicit.* Especially in data tasks, being explicit about your intentions is critical for both defensive practices and maintenance. Debugging a production pipeline is easier when you know exactly what the intent was.

- *Simple is better than complex. Complex is better than complicated.* Best practices in DAG design are to have many simple tasks that then combine to create a complex workflow. This generally means the workflows handle error recovery and retries better, are easier to extend in the future, and are easier to debug when failures do occur.

- *Special cases aren't special enough to break the rules. Although practicality beats purity.* Consistency is key to your code base for maintainability's sake, so where possible, you should make sure that you're doing things in a consistent way. However, engineering organizations do need to get things done, so you should be willing to embrace edge cases during both design and implementation.

- *Errors should never pass silently. Unless explicitly silenced.* This is a special case of explicit versus implicit. Especially in data work, errors should be explicitly handled either through an exception or logged for follow-up.

- *Now is better than never. Although never is often better than *right* now.* When building DAGs and your broader operational architecture, there is a natural tendency to start "future-proofing" the process or over-abstracting immediately – don't. You should take great care as you choose to implement your interfaces and systems; after all, you'll be responsible for maintaining them, and the effort put in now could potentially be used somewhere else.

- *If the implementation is hard to explain, it's a bad idea. If the implementation is easy to explain, it may be a good idea.* This is another case of simple/complex/complicated, but one that should especially be useful during the design phase. If you can't explain something simply enough, it probably isn't a good idea to implement it – after all, data is ever changing and evolving and your systems will have to evolve with it.

Idempotency

Idempotency describes an operation that can be applied multiple times without changing the result after an initial application. It is our experience that the biggest operational issues can be avoided if this concept is considered first and foremost during all phases of design for tasks in Airflow.

Code as configuration

Code as configuration refers to a software design principle where configuration settings and parameters are stored as code separately from the executing code base. This allows configuration parameters to be defined as code and integrated directly into the code base using programming constructs such as variables, classes, or functions. This allows developers to manage and modify application behavior in a more systematic and version-controlled manner. This approach offers benefits by increasing consistency, easier collaboration, and more ergonomic workflows that allow the application of common software development and operations paradigms.

This paradigm is very powerful and does allow you to treat your configurations as though they are software; however, developers should keep in mind that the code is intended to describe discrete configuration units, and some "obvious" software patterns can result in unintended (or undocumented) dependencies if care isn't taken to ensure that your configurations have clear boundaries in the code bases that describe your workflows.

Now that we have taken a moment to find our own zen when authoring Python and shared some key insights on when you should choose Apache Airflow, let's dive into the skills most needed.

Skills to use Apache Airflow effectively

It isn't sufficient to just know Airflow; there are other skills that are required and will impact your team's effectiveness at providing value with Airflow. As you're reading this book, think about your teams' skills in the following areas to understand if you need additional support or capabilities to make the most out of your platform:

- **Python**: DAGs and plugins are written in Python. As such, in order for Airflow to be effectively adopted, you should be able to write and understand Python code.

- **Application testing**: Your company should have relatively mature testing processes and practices in place to ensure that the plugins you write and the DAGs you author can be reasonably expected to function as intended. The team responsible for the health of your Airflow instance(s) and plugins should be able to help teams prepare for upgrades, and the teams responsible for writing workflows should be able to run automated tests prior to production deployments.

- **Domain expertise**: Knowing your data domain is the most critical skill for your success in adopting Airflow. No matter how much technical skill you have, if you don't understand the business domain you're working with, you won't be successful in the short or long run.

- **Application monitoring/alerting**: You should have relatively mature observing, monitoring, and alerting capabilities available to you in order to make the most out of Airflow. Some common activities are things such as monitoring for application activity (that is, health checks), sending alerts to on-call team members, and automated messages about status.

These skills identified are key to making the most out of Apache Airflow and should be reviewed periodically to see how you are progressing and how you can improve in different areas.

Summary

In this chapter, we spent time learning the basics of what data orchestration is and what problems companies and engineers are facing today. In addition, we introduced Apache Airflow, the leading data orchestration and workflow management tool. We also covered what you can expect over the course of this book. It is important to remember that Apache Airflow requires multiple baseline tools and knowledge areas to be most successful. Although these areas are needed for best use, each is a learnable topic and can be picked up at a quick pace.

At the core of Airflow use is Python code. To be the best data engineer using Airflow, you need to understand the core concepts of Python code and how it will orchestrate your stack of data tools. Taking time to review these core concepts and understand the use cases that are being tackled by Airflow will lead to scalable systems of code and optimization opportunities.

In the next chapter, we will introduce the basics of DAGs and tasks. We cover new tips around task decorators and organizing your task groups, as well as walking through an example. You can expect to get your first Airflow DAG running and operational.

2

Core Airflow Concepts

Apache Airflow is built upon the main concepts that will be discussed in this chapter. At the heart of Airflow are the main concepts that streamline the process of defining, executing, and monitoring tasks. These concepts include tasks, task groups, and triggers. Each of these constitutes **directed acyclic graphs (DAGs)** and allows you to take advantage of Airflow. Understanding each of these building blocks is essential for leveraging Airflow's full potential at scale and ensuring that workflows are automated and optimized.

In addition to the basics of DAGs, this chapter will introduce the concept of a command-line interface to run Apache Airflow locally on a computer or a virtual machine. The process is easy to follow, and multiple tools have been created to initiate this process. The same tools that we will use to begin creating DAGs and setting up Airflow can be used in more complex situations as well, so these tools are industry standards.

In this chapter, we're going to cover the following main topics:

- Getting Apache Airflow running on your local machine with `airflowctl`
- Building blocks of DAGs
- How to best take advantage of task groups and organizing DAGs

Technical requirements

In previous chapters, we primarily covered aspects related to Apache Airflow, but we didn't cover any basics or specific use cases. Starting in this chapter and going forward, we will assume that you have an Airflow environment set up on your local machine and understand how to access it.

Installing Airflow locally requires a few different technologies and prerequisites. Specifically, we recommend installing a recent version of Python, as older versions don't have long-term support. In addition, the minimum memory required to run Airflow is 4GB, although this requirement depends on how large the deployment grows and the complexity of the DAGs.

A **command-line interface** (CLI) is a text-based user interface, used to interact with software and operating systems. Through a CLI, users input text commands to perform specific tasks or operations, directly communicating with a system. This interface allows efficient execution of commands, automation of tasks through scripting, and access to a wide range of system functions. CLIs are especially favored by developers and system administrators for their precision, scriptability, and low resource consumption compared to **graphical user interfaces** (GUIs).

The `airflowctl` CLI is a CLI tool specifically designed for interacting with Apache Airflow environments. It allows users to manage and control various aspects of their Airflow deployments directly from the command line. With `airflowctl`, users can execute tasks such as triggering DAGs, pausing or unpausing them, creating or listing connections, and accessing logs. This tool simplifies the process of managing Airflow workflows, enabling efficient operations and monitoring of tasks within an Airflow instance. We will step into more complex ideas in later chapters on use cases for the Airflow CLIs.

The easiest and faster way to get started with Airflow is to use the Airflow CLI command within your Terminal or Command Prompt:

```
$ airflowctl
```

Running the command initially will check whether it has been previously installed on your local machine. If you receive the following prompt, follow the following installation instructions:

```
[kendrickvandoorn@Kendricks-MacBook-Pro-2 ~ % airflowctl
zsh: command not found: airflowctl
```

Figure 2.1: airflowctl setup

Using `airflowctl` (`https://github.com/kaxil/airflowctl`) allows the initial installation to be completed externally of Docker containers or Kubernetes.

It is recommended to install the CLI using `pip`. If you do not have pip installed, you can install it following these instructions: `https://pip.pypa.io/en/stable/installation/`. Using `pip`, the following command installs the CLI:

```
$ pip install airflowctl
```

If no errors are encountered during the installation, the next step is to run the initialization command. Doing so creates a project directory called `my_airflow_project` in your current folder and launches an Airflow web server:

```
$ airflowctl init my_airflow_project --build-start
```

Directly after running this initial command, you will find the project folder location and other initializing information at the top of the command terminal:

```
kendrickvandoorn@Kendricks-MacBook-Pro-2 ~ % airflowctl init my_airflow_project --build-start

Project /Users/kendrickvandoorn/my_airflow_project added to tracking.
Airflow project initialized in /Users/kendrickvandoorn/my_airflow_project
Virtual environment created at /Users/kendrickvandoorn/my_airflow_project/.venv
```

Figure 2.2: The Airflow project initialization

In this example project, the project folder is located at /Users/kendrickvandoorn/my_ airflow_project\ x and the folder has been initialized with a preset organization of folders and files, necessary to kick off this first project. Note that the name my_airflow_project from the CLI command is the same as the project folder name, but it can be adjusted to your own needs.

As the airflowctl CLI continues to run, the web server, trigger, and scheduler will come online, and their tasks will be identified at the start of each line. For example, review the following screen capture displaying an Airflow logo, which shows that the trigger and scheduler are online and the following actions are aligned to the different servers.

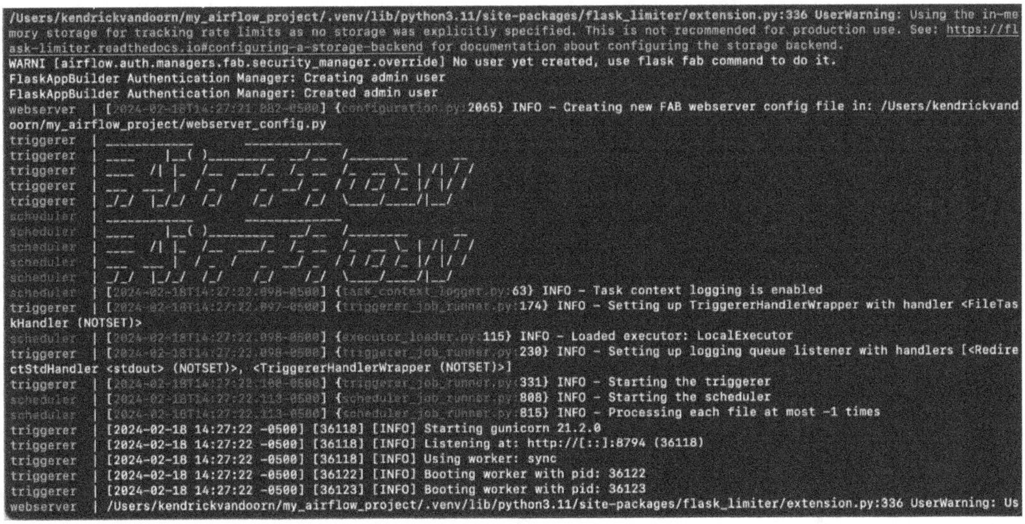

Figure 2.3: The Airflow command line

As the command line continues to run, multiple basic checks, connections, and permissions will be brought online and set. As we continue learning more about Airlfow, many of these pre-deployment decisions can be made and adjusted in different definition folders ahead of time, ensuring that the Airflow instance meets your needs relating to sizing, security, and connections. If the initialization was successful, you will see the following statements, giving details on how to access the web server and log in:

```
standalone | Airflow is ready
standalone | Login with username: admin  password:  password
standalone | Airflow Standalone is for development purposes only. Do not use this in production!
```

Figure 2.4: Airflow initialization success

Within this terminal output, you can see the `Airflow is ready` statement, showing that the process was successful and Airflow is running locally on your machine. Note the username shown is `admin`, which is default when first installing. The password provided is not default but a randomly generated password, specific to your localhost. Leave the terminal open and running to maintain the Airflow environment in your local environment.

The `airflowclt` CLI not only launches Airflow locally but also provides management of the product as it runs. All logs and information relating to Airflow will be displayed within the command terminal as it persists until turned off. Note that the different services constantly check and handle signals, which are reported in the command-line terminal.

```
triggerer | [2024-02-18T14:28:22.167-0500] {triggerer_job_runner.py:481} INFO - 0 triggers currently running
webserver | [2024-02-18 14:29:21 -0500] [36116] [INFO] Handling signal: winch
triggerer | [2024-02-18 14:29:21 -0500] [36118] [INFO] Handling signal: winch
scheduler | [2024-02-18 14:29:21 -0500] [36120] [INFO] Handling signal: winch
webserver | [2024-02-18 14:29:21 -0500] [36116] [INFO] Handling signal: winch
scheduler | [2024-02-18 14:29:21 -0500] [36120] [INFO] Handling signal: winch
triggerer | [2024-02-18 14:29:21 -0500] [36118] [INFO] Handling signal: winch
webserver | [2024-02-18 14:29:21 -0500] [36116] [INFO] Handling signal: winch
triggerer | [2024-02-18 14:29:21 -0500] [36118] [INFO] Handling signal: winch
scheduler | [2024-02-18 14:29:21 -0500] [36120] [INFO] Handling signal: winch
scheduler | [2024-02-18 14:29:21 -0500] [36120] [INFO] Handling signal: winch
triggerer | [2024-02-18 14:29:21 -0500] [36118] [INFO] Handling signal: winch
```

Figure 2.5: Web server signals

To navigate to the web server, open a web browser, and in the URL, navigate to the following:

```
http://localhost:8080
```

Navigating to the localhost with the Airflow web server still running in the terminal, you will be prompted to sign in with the username and password. Use the username of `admin` and the password provided to you in the terminal prompt.

Completing the sign-in process will take you to the home page, which should look similar to this:

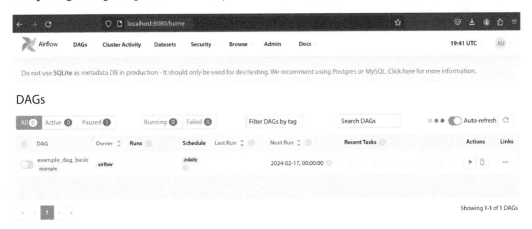

Figure 2.6: The DAG home page

In later chapters, we will explore specific details of the important sections within the Apache Airflow UI console and show examples of common management use cases.

With the Airflow environment up and running, let's take a look at the core airflow components with the DAG example that is provided.

DAGs

DAGs are the primary way to orchestrate data pipelines. They are built from Python and use a wide variety of supporting libraries. Following the previous steps to initialize a local development environment has loaded `example_dag_basic`, which we will review in this section.

Fundamentally, DAGs are comprised of tasks, operators, and sensors. There are new techniques that have been introduced over the years, relating to task groups and deferrable operators, that we will also cover in this chapter.

To access a visualization of this DAG, simply select the DAG name on the web server console. You will initially be prompted with specific information about the DAG configuration and the status of its previous runs.

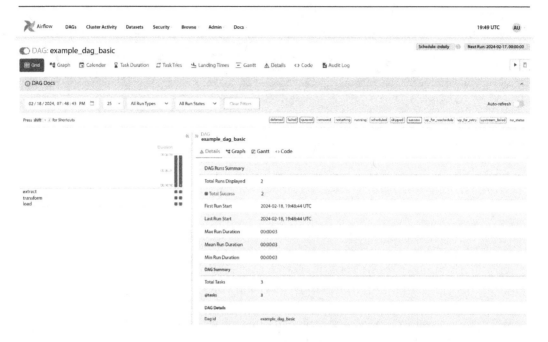

Figure 2.7: Configuring the DAG

To view a visualized representation of this initial DAG example, select the **Graph** option from the list of options at the top of the DAG area. This will bring up a graphical representation of the DAG example. It is important to note that this DAG has three tasks that run sequentially. In more complex DAGs, there will be tasks that run in parallel and some that wait for a specific trigger, and they may be too complex to visualize. It is always recommended to break up complex, monolithic software if possible.

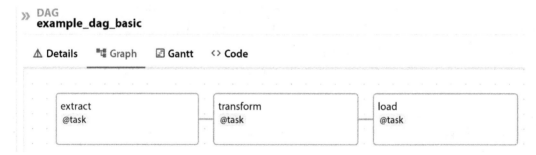

Figure 2.8: Trigger mapping

If `example_dag_basic` is not included in your initialization, you can find the full block of working code at the end of this chapter or navigate to the course book's GitHub page.

This DAG example's goal is to complete a simple extract, transform, and load function of data. Ultimately, we will show the expected output and how to confirm completion.

Decorators and a DAG definition

The first step with any Python file is to import the required libraries and support functions. An Airflow DAG is no different. We begin by importing JSON and Airflow decorators:

```
import json

from airflow.decorators import (
    dag,
    task,
)
```

Importing the DAG and task decorators allows us to declare the DAG more easily than previously. This switch removes the need to declare the "with DAG as" function, as previously required. Let's declare the DAG and identify some of the fields that are required, such as `schedule`:

```
@dag(
    schedule="@daily",
    start_date=datetime(2023,1,1),
    catchup=False,
    default_args={
        "retries":2,
    },
    tags=["example"],
)
```

In the preceding code excerpt, we declared the schedule interval, `start_date`, `catchup`, `default_args`, and a tag. Let's take a quick look at each.

Scheduling with Apache Airflow and moving away from CRON

The schedule interval can be defined by a set date time that Airflow stores or via Cron jobs. View the following table for further information on how the schedule intervals complete. Each of the times identified is set as UTC, so keep this in mind as you schedule your workloads to complete.

Date time	Interval
@none	Manual or a trigger required to execute the DAG
@hourly	Executes at the beginning of each hour
@daily	Executes at midnight every day
@weekly	Executes at midnight every Sunday
@montly	Executes at midnight on the first day of each month
@yearly	Executes on January 1 at midnight every year

Table 2.1: Schedule intervals for the date time

Next, `start_date` is declared. It is important to declare the start date, as you may want the pipeline to execute on the same day, in the future, or in the past. If `start_date` is scheduled in the past and `catchup` is set to `True`, then the DAG will rerun the number of times in accordance with the schedule. In the example given, `catchup` is `False`, so there will be no DAG runs for the previous days since January 1, 2023. If we were to set catchup equal to `True`, then we would expect Airflow to execute the number of jobs equal to the number of days between January 1, 2023, and the present.

Default arguments are default settings that we want to adjust. In this example, we alter the retries to be equal to 2. This is the number of times the DAG will retry tasks if they fail. This is a useful tool if the DAG is attempting to reach a database that often has connection or update issues, as it will re-attempt to complete the tasks at a later time without affecting the other tasks.

Finally, we apply a tag to the DAGs so that we can begin grouping them. The tag is shown in the Airflow UI here and can be useful for tagging different DAGs for different systems, teams, or users in the future.

Tasks

Tasks make up a DAG and move in order of completion, according to how they are defined. Tasks are often displayed in block formats to show how they will execute. In this basic example, there are three tasks to be completed in order. Tasks are also the most basic unit of execution in Airflow. A task represents a single operation or job that needs to be performed. Tasks are defined by operators in Airflow, with each operator determining the type of task performed. Common tasks include executing Python functions, running database queries, or performing data transformations.

Before we jump into analyzing this basic DAG example and the tasks associated with it, we need to explore operators and how they are used.

Task operators

In Apache Airflow, an operator represents a single, idempotent task that is part of a workflow defined by a DAG. Each operator in Airflow specifies a particular type of work to be done, encapsulating the logic to perform a specific task. Each operator is designed to perform a distinct function, such as running a Python function, executing a SQL query, transferring data, or even calling an external system.

Airflow comes with a large set of operators to choose from for different tasks, enabling users to start with a wide range of tasks out of the box. Some of the commonly used types include the following:

- `BashOperator`: Executes a bash command
- `PythonOperator`: Runs a Python function
- `SqlOperator`: Executes a SQL command
- `DockerOperator`: Runs a Docker container
- `HttpOperator`: Sends an HTTP request

Operators can also be customized and extended to fit specific requirements, offering flexibility in how tasks are executed within Airflow DAGs. This design allows users to create complex data pipelines that are understandable, maintainable, and scalable. Most operators are open source and available for review by searching for their names on GitHub.

The first task – defining the DAG and extract

We will form three tasks to complete this basic DAG example. An extract, transform, and load task will each complete a different function. The first task is the extract task. The following code snippet is pulled from the basic DAG example:

```
def example_dag_basic():
    @task()
    def extract():
        data_string = '{"1001": 301.27, "1002": 433.21,\
                        "1003":502.22}'
        order_data_dict = json.loads(data_string)
        return order_data_dict
```

We begin by declaring @task to identify whether the following function is to be a task utilized by Airflow. In this first task, we define it as extract, not accepting any arguments or variables. Inside extract(), we define a simple data_string that is then loaded into a data dictionary. Finally, the information is returned at the end of the function.

In more complex situations and examples, we will extract data from APIs, data lakes, and data warehouses. Creating a connection and preparing an extraction are similar processes.

Defining the transform task

In the next task, we will perform a simple transformation action. This takes in the data dictionary and computes the total order value:

```
@task(multiple_outputs=True)
def transform(order_data_dict: dict):
    total_order_value = 0

    for value in order_data_dict.values():
        total_order_value += value

    return {"total_order_value": total_order_value}
```

The transform task accepts `order_data_dict` as an input to complete the transformation task. Next, we set the total order value variable equal to 0 to establish it for use in the next loop. A `for` loop is created to calculate the total order value by summing the values. Finally, we return the total order value for use in the next task.

In this task, we can see a new argument passed into `@task`, `multiple_outputs`. Completing this unrolls the data dictionary into separate XCom values.

Xcoms

XComs, short for **Cross-Communications**, are a commonly used feature in Apache Airflow. They are designed to facilitate the sharing of data between tasks within a DAG. This key idea enables tasks to exchange messages or data snippets such as task states, return values, or any other execution-related information. This can be extremely valuable and useful when it comes to organizing tasks to run after a certain task is confirmed as completed or failed.

XComs work by allowing one task to push data to Airflow's metadata database, where it is stored under a specific key. Another task can then pull this data using the key, allowing data to be passed between tasks even though they may execute on different workers or at different times. This feature is particularly useful for tasks that have dependencies on the outcomes or output of preceding tasks.

The use of XComs promotes decoupling tasks within a workflow, enhancing the modularity and reusability of a DAG's components. However, it's recommended to use XComs judiciously, as they are intended for small amounts of data. Large data transfers are better handled by external data storage systems or services, with Airflow managing the orchestration of tasks rather than the movement of significant data volumes.

Defining the load task

In the final of the three steps, we will complete the loading of the transformed data. In this example, we simply output the total order value via a `print` function so that we are able to confirm its success in the logs:

```
@task()
def load(order_data_dict: float):
    print(f"Total order value is: {total_order_value:.2f}")
```

In the preceding code, we limit the loading of data to a simple example of printing to showcase how the task is completed. In more complex examples, we would expect to load the transformed data into a data warehouse or another system for use.

Setting the flow of tasks and dependencies

Finally, we will need to set the order of the tasks. In this example, we declare them in order using variables. In other examples we cover, you may see the use of the >> symbol, which declares that the task on the left precedes the task on the right. An example of this can be as follows:

```
extract >> transform >> load
```

This is not how the identified DAG example performs organization, and we recommend following the DAG example as follows:

```
    order_data = extract()
    order_summary = transform(order_data)
    load(order_summary["total_order_value"])

example_dag_basic()
```

As you can see from the code excerpt, we set the extract() function to be callable by order_data, which is then passed through transform(). Then, transform is passed through load() via the order_summary variable. Setting this order effects the expected ordering of the tasks.

Last, but most important, is to call the example_dag_basic() function at the end of the code. If this step is not complete, then the DAG will never operate.

Executing the DAG example

To execute the DAG, you will need to navigate back to the Airflow UI. Within the UI, follow these steps to execute the basic DAG example and confirm the completion of the extract, transform, and load example:

1. Navigate to http://localhost:8080.
2. Log in as admin with the associated password.
3. Confirm that the DAG example is on the UI.
4. Under **Actions** on the far right, click the green play button to execute the DAG.

Another way to accomplish this is to click the slider to the left of the DAG name, which will turn on the DAG and begin running it at the identified schedule.

It may take the DAG up to 30 minutes to complete the tasks. Once complete, we will follow these steps to confirm completion:

1. Click on the DAG name example_dag_basic to navigate to the DAG overview.
2. Select **Graph**.

3. Select the load task by clicking on the task box.

4. Select **Logs** from the menu bar.

5. Confirm that the total order value is 1236.70.

Task groups

To manage the complexity and enhance the readability of DAGs, Airflow 2 introduced task groups. Task groups allow you to organize tasks into hierarchically grouped subsets within a DAG. This organization is not only beneficial for visual simplification in the Airflow UI but also for logical partitioning, making large workflows more manageable and understandable.

As DAGs become more complex, especially large-scale enterprise structures and models, they become harder to understand. These groupings allow the visual grouping of tasks within the Airflow UI, allowing smaller overviews that show the different sections of the DAG.

A common example of task groupings occurs when a team has multiple transformations or extraction steps for their DAG, and they want to visualize the section as simply as possible.

If the preceding basic DAG example were to include two separate sources of data, we could create tasks for each of these extractions and group them together. An example set of code would be the following:

```
@task_group(group_id='extraction_task_group')
def tg1():
    t1 = extract()
    t2 = EmptyOperator(task_id='task_2')


    t1 >> t2
```

In this example, we defined the task group as extraction_task_group and created two separate tasks. The t1 task uses the original extract() task, and the t2 task uses the EmptyOperator, which does nothing. We set the t1 task to occur before the t2 task.

We can amend the order of tasks being completed to reflect the new task group as the starting point. The following diagram reflects the new visualized task order:

Figure 2.9: Updated task order

To expand the task group, simply select + 2 tasks on the task group and expand the section to view the tasks within:

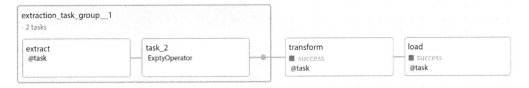

Figure 2.10: Expanded view of new task

Furthermore, Apache Airflow supports nested task groups, allowing users to further organize and structure workflows with greater granularity. Nested task groups enable you to create hierarchical structures within DAGs, where a task group can contain other task groups. This hierarchical organization also aids in debugging and maintaining a workflow, as it encapsulates related tasks into distinct, manageable units that can be developed, tested, and monitored independently.

Triggers

Triggers determine the conditions under which a task or DAG is executed. Airflow provides flexible triggering mechanisms, allowing tasks to run based on time schedules (e.g., daily and hourly), external events, or the completion of other tasks. Understanding triggers is crucial for scheduling tasks in a way that aligns with your operational requirements and dependencies. Earlier, we covered scheduling and setting intervals for the DAGs, but specific triggers within the DAG and tasks can adjust the timing of actions.

An example of a trigger in Apache Airflow is the `TimeDeltaTrigger`, which schedules tasks to run at a specific interval of time after the completion of another task. This trigger is part of Airflow's dynamic task mapping and deferred operator capabilities, allowing workflows to dynamically adjust based on conditions at runtime.

For instance, suppose in the basic DAG example that we decide that the initial transform task needs to run 30 minutes after the first extraction task group completes successfully. You could use `TimeDeltaTrigger` to achieve this by setting it to delay the execution of the aggregation task by 30 minutes.

This is accomplished by adding a trigger to the task definition in the form of the following:

```
trigger=TimeDeltaTrigger(timedelta(minutes=30)),
```

Common types of triggers used by teams include time-based, dependency-based, and event-based triggers. Let's briefly explain each.

Time-based triggers include both CRON expressions and interval-based scheduling. Cron expressions are one of the most common triggers, allowing tasks to be scheduled at regular intervals, specified by cron syntax (e.g., 0 0 * * * for a task that runs daily at midnight). Interval-based scheduling is where tasks can be scheduled to run at fixed intervals of time (e.g., hourly or daily), using a predefined schedule interval.

Dependency-based triggers are comprised of upstream task completion, an external task sensor, and many other options. Upstream task completion is where a task can be triggered when one or more specified upstream tasks have been successfully completed. This is crucial for workflows where tasks have direct dependencies on the output or success of other tasks. The external task sensor trigger waits for a specific task in another DAG to complete before proceeding. It's useful for coordinating tasks across different workflows.

The last of the most common triggers are event-based triggers, which are Webhooks and sensors relating to email or other notification services. Webhook trigger tasks can be triggered by external events via Webhooks. This is useful for workflows that need to start based on actions or signals from external systems. An email sensor triggers a task based on the arrival of an email that meets specific criteria, which is useful for workflows that initiate in response to email notifications.

Other triggers worthy of note are data availability triggers and manual triggers, which can be triggered by manual user actions.

Summary

In this chapter, we introduced the basics of DAGs, tasks, operators, Xcoms, task groups, triggers, and the `airflowctl` CLI. We covered interacting with the Airflow console and reviewing the basics of DAG authoring, with a typical example included in the initial locally run version of Airflow.

Each of these topics is critical to understand and master before attempting to create large ETL pipelines or other ML/AI use cases with Airflow. It is recommended that you spend time reviewing the basic DAG example and practice the different triggers or operators identified in this chapter, ensuring that you feel comfortable building larger-scale systems as you grow as an engineer.

In the next few chapters, we will expand on this initial basic DAG example with a real ETL pipeline and perform our first data extraction.

Part 2:
Airflow Basics

This part has the following chapters:

3

Components of Airflow

Apache Airflow is a distributed system with multiple components. While distributed systems are inherently complex, the components themselves are relatively simple. It is important to understand each component's specific role and the role they play in the operation of an Airflow application. Understanding their setup, operation, and maintenance will help you scale confidently and become experts at operating your Airflow environments in production. In this chapter, you will learn about each component's responsibility and capabilities within the overall application, how to select certain configurations for certain components, and how to determine which capabilities you will need in order to achieve your business goals.

It is important to focus on understanding basic components such as these as there is often optimization and latent capacity opportunities to be found as the total number of tasks and jobs that Airflow orchestrates grows.

In this chapter we are going to cover the following main topics:

- Overall architecture and key components
- Which Executor is right for different use cases
- Closer look at optimizing the Scheduler

Technical requirements

Similar to previous chapters, we expect that you have an Apache Airflow environment set up on your local machine and understand how to access it, whether via the user interface or through a command line interface. If you have not completed these steps, we recommend referencing previous chapters or navigating to the Quick Start guide maintained by the open-source community for the most up to date process.

A general understanding of concepts such as distributed software architecture, Kubernetes, and system design is needed to get the most out of this information.

Overall architecture

Apache Airflow at its core is a set of components working together, allowing you to build and run workflows or Directed Acyclic Graphs (DAGs). These workflows run on top of several microservices that coordinate workers to execute tasks on a specified schedule.

Apache Airflow's architecture comprises several key components that work together to orchestrate data pipelines efficiently. The primary components include:

- **Metadata Database**: Stores metadata related to DAG runs, task instance status, and other key metadata. It allows your Airflow instance to keep track of task states, DAG versions, and offers persistence.

- **Scheduler**: Responsible for triggering the task instances, based on assigned time or an external trigger. It will check the DAGs constantly to see if they can be triggered.

- **Triggerer**: responsible for storing and executing asynchronous functions that are instantiated from Trigger classes.

- **Executor**: Determines how tasks are going to be executed in Airflow.

- **Workers**: Pick up the tasks that are scheduled for execution and are responsible for actually "doing the work."

- **DAG Directory**: Where Airflow searches for Python files containing DAG definitions. Mounted by many components interfaces to the file system.

- **Web Server**: Provides the web-based user interface for Apache Airflow.

- **User Interface**: From the webserver, the UI allows users to monitor DAGs, view successful and failed task runs, inspect logs, and manage the Airflow environment.

Each of these components serves a unique purpose in the Airflow ecosystem, collectively enabling smooth orchestration and management of complex data workflows. Understanding these core elements provides a solid foundation for working with Apache Airflow's architecture.

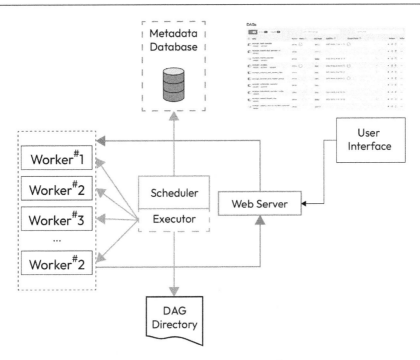

Figure 3.1: Overall architecture of Apache Airflow Components

The webserver, scheduler, and executor are Airflow processes. The Metadata Database or Database is a separate service required to be provided to Airflow for storing metadata from the webserver and scheduler. The DAG Directory or `dags` folder must be accessible by the scheduler and is often included in the same working directory.

The webserver visually displays information about the current status of DAGs and pipelines as well as provides the user the ability to review key information in different views and manually trigger DAGs. The scheduler serves to parse DAG files from the DAG directory and determine the tasks to run and places them in a queue.

To execute these tasks in the queue, there are multiple options to select from depending on the business goals and requirements. Apache Airflow can be installed and run in various ways such as on a local machine, a single machine, or on a distributed network of multiple machines. Each of these approaches brings different benefits and levels of complexity and requires a different executor.

Executors

Executors determine how task instances are going to be executed in an Airflow environment. They are *pluggable*, allowing teams to swap different executors based on their business goals and needs. Each Airflow environment can only have one executor configured at a time which can be adjusted within the configuration file.

You may see the executors displayed as `SequentialExecutor` in the official documentation and in the configuration files. In this chapter we will keep the words separated for ease of documentation and refer to them by their type such as Sequential instead of `SequentialExecutor`.

As of the time of writing there are multiple types of executors, and the community continues to expand on options. Table 4.1 contains a list of executors currently available, as well as ones that have been deprecated. In the following table, we will cover additional details around the most common executors and best use cases.

Executor	Remote	Parallelism	Installation and Maintenance Burden	User Cases
SequentialExecutor	No	No	Very Easy	Demo/Testing
LocalExecutor	No	Yes	Easy	Single Machine or small Development environment
CeleryExecutor	Yes	Yes	Moderate	Scaling across multiple machines and workers
CeleryKubernetes Executor	Yes	Yes	Complex	Same as CeleryExecutor, but handles high loads at peak time and runtime isolation of KubernetesExecutor
Dask Executor	Yes	Yes	Moderate	Parallel computing across distributed architecture
Kubernetes Executor	Yes	Yes	Complex	Scale and run each task instance in own pod on Kubernetes cluster
LocalKubernetes Executor	Yes	Yes	Complex	Advantage of Kubernetes Executor with capability to run tasks with LocalExecutor within the scheduler server

Table 3.1: A description of some of the common executors currently available for executing tasks, their complexity, and operational use cases for each.

Let's jump into examples of each type and take a look at the best applications of each.

Local Executors (Sequential and Local)

If you followed the quick start instructions to install Apache Airflow and you have not made any changes, the default executor is set to Sequential Executor. This is the only executor that can be used with SQLite as SQLite does not support multiple connections.

The Sequential Executor operates as its name suggests, tasks are performed in sequence, following a logical order. Local installations of Apache Airflow often have one worker as the Sequential Executor limits one task to be executed at a time.

Figure 3.2: Set of three tasks from the basic DAG example

Looking back at the basic DAG example covered in the previous chapter, the Sequential Executor is a perfect use case. The DAG is not complex, and each task needs to be completed in order prior to moving onto the next task. The Sequential Executor is a great tool for running example DAGs and running workloads on a local machine. As the number of required workers increases, local machine performance will be slowed due to the utilization of resources.

A step up from the Sequential Executor is the Local Executor, which runs task instances in parallel on the same machine where the scheduler is running. It uses Python's multiprocessing module to spawn multiple processes, allowing for parallelism. To make the change from Sequential Executor to the Local Executor the change must be updated in the configuration file (`airflow.cfg`) by setting the executor field to Local Executor.

```
# within the config file
executor = LocalExecutor
```

If you are interested in checking which executor is being utilized by the Airflow environment that you are viewing, you can run the following command within the command line interface:

```
$ airflow config get-value core executor
```

The primary Use Cases of the Local Executor are:

- **Development Environment**: Because of its simplicity and no external dependencies, the Local Executor is commonly used in dev environments. Developers can confidently run DAGs and tasks without needing the overhead of more complex executors.

- **Small to Medium Workloads**: For production environments with limited parallelism requirements (less than five concurrent DAGs/tasks on average) or weak SLAs, the Local Executor may be a sufficient solution to the team's needs.

The Local Executor offers multiple benefits compared to the Sequential Executor and Remote Executors:

- **Parallelism**: This executor introduces the ability to run multiple task instances concurrently. The degree of parallelism essentially dictates how many task instances can be executed simultaneously at any given time. This tool is extremely beneficial when it comes to long running tasks. Parallelism is a configuration that can be adjusted to the highest limit a machine hosting the Local Executor can support.

- **Simplicity**: The Local Executor is straightforward and does not require setting up additional infrastructure components, such as message brokers for Celery Executor or a Kubernetes cluster for the Kubernetes Executor. We will cover both of these Executors later in this chapter. This makes it easier to configure, especially for new users or small setups.

- **Local Development**: It provides a more realistic testing environment compared to the Sequential Executor, which runs tasks one at a time. Developers can test parallel task execution without the overhead maintenance and set up of more complex executors.

- **Resource Utilization**: Since it runs tasks on the same machine as the scheduler it is suitable for scenarios where you want to fully utilize the machine's resources without distributing tasks across different nodes or containers.

- **Low Overhead**: With the absence of external systems to send tasks to for completion, there is reduced network latency, and there are no additional systems to monitor or maintain.

- **Transition to Production**: For small to medium-sized Airflow deployments, transitioning from a development environment using Local Executor to a production environment with the same executor is straightforward.

However, it's essential to note the limitations of the Local Executor. As the name suggests, tasks are executed "locally," so if workflows require significant parallelism or if there is a need to distribute the execution load across multiple machines for scalability or failover, then other executors might be more appropriate.

Parallelism

Let's take a closer look at the topic of Parallelism and how powerful it is for efficiently completing task instances. The Local Executor introduces the ability to run multiple task instances concurrently. Compared to the Sequential Executor limiting one task at a time, parallelism can drastically speed up the process. For example, in the next image we can visualize how a Sequential Executor may approach executing three separate tasks. Each task is marked as a different color and are not dependent on each other to execute.

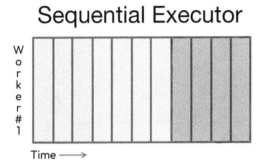

Figure 3.3: Visualization of a single worker executing three tasks across time

The Sequential Executor will take in the first task, colored green and requiring three cycles to complete. Although the second task, blue, is not dependent on the green task to begin execution it must wait for the green task to complete execution within the single worker available. The third task, purple, requires the same process of waiting for the first and the second task instances to complete prior to beginning execution. This process may be acceptable for some business cases and offers a quick way to get started.

The Local Executor offers the ability to implement multiprocessing or parallelism to complete tasks concurrently. In this following example, there are still three tasks, and each is not dependent on each other. The executor is set with two workers to complete task instances.

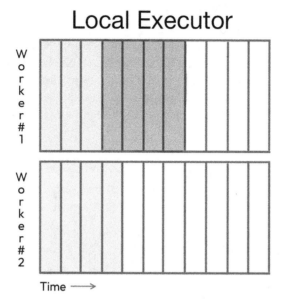

Figure 3.4: Local Executor example with parallelism

In this example, we see that both task one, green, and task two, blue, are executed at the same time as each worker independently works on them. Once task one is completed, task three, purple, can be started by Worker #1. If you are interested in more optimizations, we will cover worker pools and queues in later chapters, which can be extremely beneficial from a cost and time utilization perspective.

The degree of parallelism essentially dictates how many task instances can be executed simultaneously at any given time. This tool is extremely beneficial when it comes to long running tasks. Parallelism is a configuration that can be adjusted to the highest limit a machine hosting the Local Executor can support.

Celery Executor (Remote Executor)

The Celery Executor is one of the remote executors available in Apache Airflow. It uses Celery, a distributed task queue system that allows for executing tasks in parallel across multiple worker machines. In this setup, Celery uses a broker, such as RabbitMQ or Redis, to handle communication between the main Airflow instance and the worker machines.

Primary use cases of the Celery Executor include:

- **Scalability**: Suitable for large Airflow deployments where tasks need to be distributed across multiple machines due to the high volume of task instances or resource-intensive tasks.

- **Distributed Execution**: When you want to execute tasks on different machines with different configurations or capabilities.

- **High Availability**: With multiple worker nodes, if one worker fails, others can still process the tasks, ensuring that the system is resilient to failures.

- **Resource Segregation**: If certain tasks require specific system resources or configurations, they can be directed to designated workers set up for those needs.

The Celery Executor offers multiple benefits compared to the local and remote executors:

- **Horizontal Scalability**: As the workload grows, you can simply add more worker nodes to handle the increased load without modifying the existing infrastructure.

- **Flexibility**: You can configure different worker nodes for different types of tasks based on resource requirements, ensuring optimal resource utilization.

- **Decoupling**: The decoupled architecture means that the web server and scheduler components of Airflow are separate from the task execution machines or workers. This separation ensures that resource-intensive tasks won't slow down or affect the performance of the scheduler or webserver / user interface.

- **Concurrency**: By distributing tasks across multiple machines, you can achieve a high degree of parallelism, allowing many tasks to run concurrently, thus reducing overall execution time for large workflows.

As you begin to look at the different executors, the complexity of set up and management increases from the basic Sequential Executor. For the Celery Executor it is best to keep the following in mind:

- **Setup Complexity**: Compared to LocalExecutor or SequentialExecutor, setting up CeleryExecutor requires additional components like a message broker (RabbitMQ or Redis) and result backend, making the initial setup more complex.

- **Operational Overhead**: Monitoring and maintaining multiple components (Airflow, Celery, Workers) can introduce operational challenges.

- **Latency**: There might be slight latency introduced due to messages being passed between the main Airflow instance, the broker, and the worker nodes.

- **Broker Limitations**: The choice of message broker comes with its own set of limitations, quirks, and maintenance challenges.

- **Cost**: Running multiple worker nodes might increase infrastructure costs, especially if they are underutilized.

- **Version Synchronization**: Ensuring that all worker nodes are running the same version of Airflow, and all dependencies can be challenging in a distributed setup.

While the Celery Executor provides scalability and distributed execution capabilities essential for large-scale Airflow deployments, it comes with increased complexity and operational challenges. It's essential to weigh the benefits against the limitations based on the specific requirements of your workflows.

Kubernetes Executor (Remote Executor)

The Kubernetes Executor is a remote and dynamic executor for Apache Airflow that runs task instances in individual Kubernetes pods. This executor was introduced to leverage the capabilities of Kubernetes, enabling on-demand spawning of pods for task execution and meet the growing need of enterprise teams with highly complex and large workloads.

Primary use cases of the Kubernetes Executor include:

- **Running tasks that require access to large amounts of resources:** The KubernetesExecutor can be used to run tasks that require access to large amounts of resources, such as CPU, memory, and GPU resources. This is because Kubernetes can dynamically allocate resources to pods as needed.

- **Running tasks that need to be executed in a specific environment:** The KubernetesExecutor can be used to run tasks that need to be executed in a specific environment, such as a specific version of Python or a specific set of libraries. This is because Kubernetes can run pods with different containers, each of which can contain a different environment.

- **Running tasks that need to be fault-tolerant:** The KubernetesExecutor can be used to run tasks that need to be fault-tolerant. This is because Kubernetes can restart failed pods and reschedule tasks to other pods.

- **Scalability**: The executor is suitable for environments where the task load varies significantly. Kubernetes can quickly scale up or down based on demand.

- **Distributed Execution**: Useful when tasks need to be distributed across a Kubernetes cluster, whether in a cloud environment or on-premises.

- **Ephemeral Environments**: For tasks that require a clean environment every time they run, spawning a new pod provides an ephemeral setting.

With the introduction of the Kubernetes Executor there have been multiple benefits identified compared to the local and other remote executors:

- **Dynamic Scalability**: Unlike static configurations where resources are pre-allocated, with Kubernetes Executor, resources are allocated only when tasks need to run, optimizing resource usage.

- **Integration with Cloud**: Many cloud providers offer managed Kubernetes services (e.g., GKE, EKS, AKS). The Kubernetes Executor integrates with these services and can be deployed using tools such as Helm, providing cloud-native scalability and management.

- **Failover & Redundancy**: Kubernetes inherently provides features like auto-restarts, replacing failed pods, and spreading pods across nodes, which increases the reliability of task execution.

- **Customization**: Each pod can be customized using Kubernetes primitives. This allows for setting specific configurations, secrets, volumes, and other necessities for tasks on a case-by-case basis.

With this additional capability and control over the task instances and the worker that completes the execution the complexity of maintenance grows. For the Kubernetes Executor it is best to keep the following in mind:

- **Setup Complexity**: Deploying and managing a Kubernetes cluster, especially if not already in use, can be complex and requires expertise in Kubernetes, cloud architecture, and distributed networks.

- **Overhead**: For very lightweight or quick tasks, the overhead of spinning up a new pod might be significant compared to the actual task runtime.

- **Cost**: While dynamic scaling can be cost-effective, there's still the underlying cost of maintaining a Kubernetes cluster. Additionally, frequent pod creation and destruction could lead to increased costs or cost savings in some instances.

- **Persistent Data**: Pods are ephemeral, and storing persistent data can be challenging. While there are ways around this using persistent volumes, it introduces additional complexity.

- **Learning Curve**: For teams unfamiliar with Kubernetes, there might be a steep learning curve both in terms of understanding Kubernetes concepts and debugging issues specific to the platform.

- **Network Latency**: Spinning up pods might introduce network latency, especially if images need to be pulled from a registry or if there are initial setup tasks in the pod.

- **Startup Latency**: Images have to be pulled from the registry (if not already cached) and containers go through a startup process for every task. Depending on the image architecture, the startup time on the container can be considerable.

- **Version Compatibility**: Ensuring compatibility between Airflow versions and Kubernetes versions, as well as keeping up with Kubernetes API changes, can be a maintenance concern.

In summary, the KubernetesExecutor offers a highly dynamic and scalable execution environment for Airflow tasks, capitalizing on the strengths of Kubernetes. However, it may introduce complexity and overhead, especially for teams unfamiliar with Kubernetes or for workflows with lightweight tasks. As always, the choice of executor should be based on the specific requirements and context of the deployment.

Dask Executor (Remote Executor)

The DaskExecutor is an executor for Apache Airflow that leverages Dask, a flexible parallel computing library for analytic computing. Dask can be used to build parallel, distributed computation frameworks that scale from a single machine to a cluster of machines. When used in Airflow, the DaskExecutor dispatches task execution to a Dask cluster.

Primary use cases of the Dask Executor include:

- **Machine learning:** The Dask executor is well-suited for machine learning tasks, such as training and evaluating models. This is because Dask can distribute these tasks across multiple workers, which can significantly speed up the training process.

- **Data science:** The Dask executor can also be used for data science tasks, such as data preprocessing and analysis. This is because Dask can distribute these tasks across multiple workers, which can help to improve performance and scalability.

- **Other computationally expensive tasks:** The Dask executor can also be used for other computationally expensive tasks, such as video processing and scientific computing.

The Dask Executor offers multiple benefits compared to the local and remote executors:

- **Shared Infrastructure**: For teams or organizations already invested in Dask for distributed computing, the DaskExecutor allows for using the same infrastructure for Airflow task execution, ensuring resource optimization.

- **Flexibility**: Dask provides a flexible platform that can be run on a single machine (in a multi-threaded or multi-process manner) or on a distributed cluster. This flexibility is beneficial for varying scales of deployment.

- **Python-native Ecosystem**: Dask is tightly integrated with the Python ecosystem, making it a natural fit for data-intensive workflows written in Python.

As the Dask Executor is a newer option within the community it is best to keep the following in mind:

- **Setup Complexity**: If you don't already have a Dask setup, introducing a Dask cluster and ensuring its smooth operation can be complex and there may be more value in utilizing a different executor.

- **Operational Overhead**: Managing and monitoring a Dask cluster, especially in a production environment, can introduce additional operational challenges.

- **Dependency Management**: Ensuring that all Dask workers have the correct and consistent environment and dependencies can be challenging, especially when diverse tasks with different requirements are running.

- **Potential for Resource Contention**: If the Dask cluster is being used for other computational tasks apart from Airflow, there might be resource contention, leading to performance issues.

- **Network Overhead**: Depending on how the Dask cluster is set up, there might be network overhead in dispatching tasks and retrieving results, especially if tasks are I/O bound.

In conclusion, the DaskExecutor offers an avenue for distributed task execution in Apache Airflow, especially for teams already using Dask or those running data-intensive Python workflows. However, like other distributed executors, it introduces additional complexities and potential challenges in setup, management, and performance optimization. The choice to use DaskExecutor should be based on specific workflow needs, existing infrastructure, and familiarity with Dask.

Kubernetes Local Executor (Hybrid Executor)

The Kubernetes Local executor is a newer executor for Apache Airflow that allows you to run tasks locally using the Local Executor or on Kubernetes using the Kubernetes Executor, depending on the task instance's queue. This can be useful for tasks that have different resource requirements or that need to be executed in a specific environment.

Primary use cases of the Kubernetes Local Executor include:

- *Developing and testing Airflow DAGs locally:* The Kubernetes Local executor can be used to develop and test Airflow DAGs locally without having to deploy them to a Kubernetes cluster. This can be helpful for debugging DAGs and for quickly iterating on new features.

- *Running Airflow DAGs on a Kubernetes cluster in production:* The Kubernetes Local executor can also be used to run Airflow DAGs on a Kubernetes cluster in production. This can be helpful for tasks that have high resource requirements or that need to be executed in a scalable and fault-tolerant environment.

The Kubernetes Local executor is an important addition to the list of executors as it provides flexibility in how to run Airflow tasks, depending on the task's specific needs. This can be helpful for developing, testing, and running Airflow DAGs in a variety of environments.

Here are some additional things to keep in mind when using the Kubernetes Local executor:

- Make sure that there is enough memory and CPU resources to support the number of Kubernetes pods that the Kubernetes Local Executor may need to spin up.

- Remain aware of the increased load the Kubernetes Local Executor may have on the Database.

- Be aware that using the Kubernetes Local Executor can cause tasks to run concurrently, which can lead to race conditions. To avoid this, you may need to implement synchronization mechanisms.

Scheduler

The previous sections covered how tasks are executed and the best way to enable different use cases of tasks instances to be executed. To determine when these tasks should be scheduled for execution, we need to take a closer look at the Scheduler and its multiple responsibilities:

- **DAG Parsing**: The scheduler continuously parses DAG files in the DAG Directory to look for new tasks to schedule. It determines the execution order based on dependencies set within the DAGs.

- **Heartbeat Mechanism**: The scheduler operates in a loop, often referred to as the "heartbeat", where it continually checks for tasks to run, schedules them, and then sleeps for a short duration before checking again.

- **Dynamic Task Scheduling**: Unlike traditional cron setups where jobs are fixed, the Airflow scheduler dynamically determines which tasks should run based on their dependencies and state. This allows for more complex workflows with conditional execution paths.

Some key controls of the scheduler that we have found most useful for optimizing production environments include the following:

- **Concurrency Controls**: The scheduler respects various concurrency and parallelism settings that can adjusted in the configuration file:

  - `dag_concurrency`: The number of task instances allowed to run concurrently by the scheduler for a specific DAG.

  - `parallelism`: Global parallelism, which is the total number of task instances that can run concurrently across all DAGs.

- **Handling Failures**: If a task fails, the scheduler can retry it based on the retries parameter set in the task. It respects the `retry_delay` setting, which determines the time between retries.

- **Backfill and Catch-up**: The scheduler can backfill historical data by running missed DAG runs for a specified date range. Additionally, if catch-up is set to True for a DAG, and a DAG run is missed (e.g., due to downtime), the scheduler will execute that DAG run when Airflow is back online.

- **Resource Management and Pools**: The scheduler ensures tasks are scheduled based on resource availability. Using the pools concept in Airflow, one can limit the number of concurrent tasks that use a particular resource, ensuring no resource is over-allocated. We will go into more detail about task queues and pools in later chapters.

A key attribute of the scheduler is that it is responsible for task instances up to the point that they are entered into a queue state. Once a task is placed in the queue it becomes the responsibility of the selected executor to run the task.

Summary

In summary, understanding the components of Apache Airflow, their individual roles, and how they interact is crucial for effectively setting up, operating, and optimizing an Airflow environment. This knowledge enables you to choose the right configurations, ensure efficient task execution, and scale the system to meet your business goals confidently. By mastering these elements, you'll be well-equipped to manage Airflow in production, unlocking opportunities for optimization and leveraging latent capacities as your task and job orchestration needs grow.

In the next chapter we'll get to start using these components by covering the basics of DAG authoring.

4

Basics of Airflow and DAG Authoring

In *Chapter 3*, we went through the basics of setting up Airflow on your local machine and analyzing a basic ETL example. In this chapter, we will take it a large step further with a real-world example of extracting data from an API on a schedule. You can expect to use the knowledge from the previous chapters to tackle this problem, and we will guide you along the way.

The best way to become familiar with Airflow and understand the basics is to get "hands-on" with the process. As we will be providing complex examples in the following chapters and identifying areas of improvement in the cloud for cost and efficiency savings, it is important that these early examples lead up to those in appropriate levels of complexity.

In this chapter, we are going to cover the following main topics:

- Authoring an advanced DAG example
- Learning about operators
- Presenting findings and reviewing the information

Technical requirements

Similar to previous chapters, we expect that you have an Apache Airflow environment set up on your local machine and understand how to access it. In addition, for this chapter, it may be valuable to use a local testing tool such as Jupyter Notebook to test the code if errors are encountered.

In this section, we will go into detail on how to complete the actions within Airflow. All code for this chapter can be found at https://github.com/PacktPublishing/Apache-Airflow-Best-Practices/tree/main/chapter-04.

Designing a DAG

When starting any software or data engineering project, the initial phases of designing what it is we are to build is a critical early-stage activity. Completing this step provides a clear visual representation of how different data components and processes interact, ensuring that stakeholders and team members alike have a unified understanding of the system's structure. This common understanding helps to reduce ambiguities and miscommunications, fostering a smoother development process.

Well-defined architecture diagrams aid in identifying potential bottlenecks, redundancies, and inefficiencies, allowing teams to optimize workflows and resources from the onset. As your data engineering projects grow in complexity, they often involve integration with various data sources and tools; the diagram serves as a roadmap for integration points, ensuring that all elements are compatibly and cohesively connected.

Lastly, a detailed architecture diagram is instrumental in assessing a project's scalability and resilience, ensuring that the system can accommodate future growth and withstand potential failures. In essence, such a diagram acts as a blueprint, guiding the project's direction. We will go through creating a diagram in the next section.

DAG authoring example architecture development

Gathering requirements for a data engineering ETL pipeline is an important initial step to ensure that the designed system aligns with the business needs and technical constraints. To go about it, you need to engage with stakeholders from all teams that will be affected, such as business analysts, data scientists, and source system owners. Your goal is to understand the data sources, desired data transformations, and final consumption needs. It is also essential to factor in data volume, frequency of data updates, latency requirements, and data quality needs. Technical constraints, such as available infrastructure, tools, and existing systems, should also be evaluated.

Conducting workshops or structured interviews can help capture these requirements comprehensively. We always recommend talking to all stakeholders to better document the solution they desire and align the architecture with their goals. Additionally, documenting and revisiting these requirements periodically ensures that the pipeline remains relevant and adaptive to changing business needs.

DAG example overview

During my early career, I had the opportunity to work on the IT services team that supported all programs, including the building of the James Webb Space Telescope. This telescope is the newest in NASA's toolset, allowing for high-resolution and high-sensitivity images, allowing for the visualization of objects too old, distant, or faint for the Hubble Space Telescope. Because of this, I began down a path of wanting to stay up to date on beautiful and mesmerizing photos from outer space and view any new images that NASA releases.

To retrieve these images, I frequently make use of the NASA Astronomy Picture of the Day API (`https://apod.nasa.gov/apod/astropix.html`) to gather a new image. This is a free API requiring an API key to be created but is easily accessible.

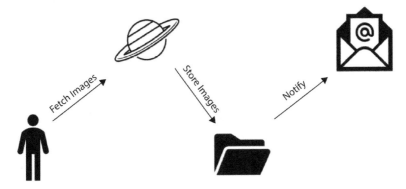

Figure 4.1: My diagram of a basic extract and notify model

Initial workflow requirements

Similar to project tracking boards in Jira, Trello, or GitHub, project requirements are broken into several sections. The goal of documenting these is to create a structured and well-designed approach. We will expand on a few key ones for this example here:

- **Functional**: The pipeline should be able to connect to the API to fetch data

- **Business**: The pipeline should be designed for scalability to accommodate growth in data volume and user activity

- **User interface**: Collect and store data for access

In future examples, it may be beneficial to stop reading for a moment and gather your own list of requirements in a separate spreadsheet or on a piece of paper. We will provide points in the book to stop and work through a problem on your own with an opportunity to compare with our own list.

Bringing our first Airflow DAG together

To bring it all together in a visualized format, there are multiple tools available. We recommend taking advantage of a tool that meets your needs and allows for the sharing of the diagram. Some quick and easy options are Microsoft PowerPoint, Google Slides, Figma, Miro, and Lucidchart. It is recommended to find a solution that allows for versioning history and general access and is available at all times. This will allow different stakeholders to review the content and offer feedback, often saving time and frustration.

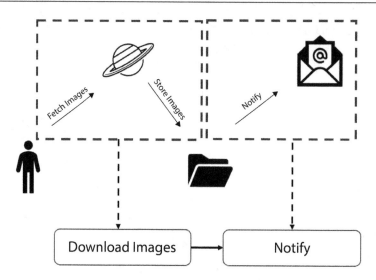

Figure 4.2: Simple architecture diagram

Let's walk through the preceding diagram to get a better understanding of what is going to happen. Similar to the example in *Chapter 3*, we plan to extract the data (which, in this case, is an image) from a source location, load it into our database (local folder), and instead of transforming, we will perform a short notification. In the next chapter, we will expand on this notification and show more options to improve this function.

A quick side note before diving into the details: It is helpful to grab the icons and logos from a quick internet search to better visualize the process of your DAGs. It helps to visualize your plans prior to starting to assist in thinking about how the architecture will flow. As the complexity of the pipeline grows, additional diagrams may be needed. These additional diagrams can include data flow, sequence, component, and deployment diagrams.

Extracting images from the NASA API

This pipeline is designed to extract an image every day, store this information in a folder, and notify you of the completion. This entire process will be orchestrated by Apache Airflow and will take advantage of the scheduler to automate the function of re-running. As stated earlier, it is helpful to spend time working through practicing this in Jupyter Notebook or another tool to ensure the API calls and connections are operating as expected and to troubleshoot any issues.

The NASA API

For this data pipeline, we will be extracting data from NASA. My favorite API is the **Astronomy Picture of the Day** (**APOD**) where a new photo is selected and displayed. You can easily change the

API to another of interest, but for this example, I recommend you stick with the APOD and explore others once completed.

A NASA API key is required to start this next step:

1. Create a NASA API key (`https://api.nasa.gov/`).

2. Input your name, email, and planned functional use of the API.

3. Navigate to your email to locate the API key information.

Generate API Key

Required fields are marked with an asterisk (*).

First Name *

Last Name *

Email *

How will you use the APIs? (optional)

Signup

This site is protected by reCAPTCHA and the Google Privacy Policy and Terms of Service apply.

Figure 4.3: NASA API Key input screenshot

By navigating to your email address, you will an email from `noreply@api.data.gov` containing your API key information. I recommend saving this in a secure location and not uploading this information to your GitHub or other repository. In the next few chapters, we will expand on how to better manage these API keys and keep them secure.

Building an API request in Jupyter Notebook

With the environment configured and the API set up, we can begin authoring a DAG to automate this process. As a reminder, most Python code can be pre-tested in a system outside of Airflow, such as Jupyter Notebook or locally. If you are running into problems, it is recommended to spend time analyzing what the code is doing and work to debug.

In Jupyter Notebook, we are going to use the following code block to represent the function of calling the API, accessing the location of the image, and then storing the image locally. We will keep this example as simple as possible and walk through each step:

```python
import requests
import json
from datetime import date
from NASA_Keys import api_key
url = f'https://api.nasa.gov/planetary/apod?api_key={api_key}'
response = requests.get(url).json()
response
today_image = response['hdurl']
r = requests.get(today_image)
with open(f'todays_image_{date.today()}.png', 'wb') as f:
    f.write(requests.get(today_image).content)
```

The preceding code snippet is normally how we recommend starting any pipeline, ensuring that the API is functional, the API key works, and the current network requirements are in place to perform the procedures. It is best to ensure that the network connections are available and that no troubleshooting alongside the information security or networking teams is required.

Here is how the code looks in our Jupyter Notebook environment:

1. We begin by importing the required libraries to support our code. These libraries include the following:

   - `requests`: A common Python library for making HTTP requests. It is an easy-to-use library that makes working with HTTP requests simple and allows for easy use of GET and POST methods.

   - `json`: This library allows you to parse JSON from strings or files into a dictionary or list.

   - `datetime`: This library provides the current `date` and `time` parameters. We will use this later on to title the image file.

   - `NASA_Keys`: This is a local file to our machine holding the `api_key` parameter. This is used in this example to keep things as simple as possible and also mask the variable.

```
In [1]: import requests
        import json
        from datetime import date
        from NASA_Keys import api_key
```

Figure 4.4: What your current Jupyter cell should look like

2. After importing the appropriate libraries and variables required, we construct a variable called `url` to house the HTTP request call including our `api_key` variable. This allows the `api_key` variable to be included in the URL while hidden by a mask. It calls `api_key` from the `NASA_Keys` file:

```
url = f'https://api.nasa.gov/planetary/apod?api_key={api_key}'
```

3. Next, we use the `requests` library to perform an HTTP `GET` method call on the URL that we have created. This calls on the API to send information for our program to interpret. Finally, we convert this information from the `GET` call into JSON format. For our own understanding and analysis of the information being sent back, we print out the response to get a view of how the dictionary is structured.

In this dictionary, it seems that there is only one level with multiple key-value pairs including `copyright`, `date`, `explanation`, `hdurl`, `media_type`, `service_version`, `title`, and `url`:

```
In [2]: url = f'https://api.nasa.gov/planetary/apod?api_key={api_key}'
        response = requests.get(url).json()
        response

Out[2]: {'copyright': 'Victor Lima',
         'date': '2023-11-25',
         'explanation': "Immersed in an eerie greenish light, this rugged little planet appears to be home to stunning wate
         r falls and an impossibly tall mountain. It's planet Earth of course. On the night of November 9 the nadir-centered
         360 degree mosaic was captured by digital camera from the Kirkjufell mountain area of western Iceland. Curtains of
         shimmering Aurora Borealis or Northern Lights provide the pale greenish illumination. The intense auroral display w
         as caused by solar activity that rocked Earth's magnetosphere in early November and produced strong geomagnetic sto
         rms. Kirkjufell mountain itself stands at the top of the stereographic projection's circular horizon. Northern hemi
         sphere skygazers will recognize the familiar stars of the Big Dipper just above Kirkjufell's peak. At lower right t
         he compact Pleiades star cluster and truly giant planet Jupiter also shine in this little planet's night sky.",
         'hdurl': 'https://apod.nasa.gov/apod/image/2311/Kirkjufell2023Nov9_2048.jpg',
         'media_type': 'image',
         'service_version': 'v1',
         'title': 'Little Planet Aurora',
         'url': 'https://apod.nasa.gov/apod/image/2311/Kirkjufell2023Nov9_1024.jpg'}
```

Figure 4.5: Response from the NASA API call

In the next step, we will utilize the `hdurl` key to access the URL associated with the high-definition astronomy image of the day. Since I am an enthusiast and want the highest quality image available, I have decided that the highest definition available meets my user needs. This is a great example of a time to determine whether your users desire or need the highest quality available or whether there is an opportunity to deliver a product that meets their needs at a lower cost or lower requirement of memory.

We store `response['hdurl']` within the `today_image` variable for use in the next step for storing the image. This storing of `hdurl` allows for manipulation of the string later on in the next step:

```
In [3]: today_image = response['hdurl']
        r = requests.get(today_image)
```

Figure 4.6: Saving the hdurl response in a variable

4. In the last step, we make use of hdurl and append date.today() to create a new name for the image each day. This is so that an image from yesterday does not have the same name as an image from today, thus reducing the risk of overwrites. There are additional ways to reduce the risk of overwrites, especially when creating an automated system, but this was chosen as the simplest option for our needs:

```
In [9]: with open(f'todays_image_{date.today()}.png', 'wb') as f:
            f.write(requests.get(today_image).content)
```

Figure 4.7: Writing the image content to a local file

5. Finally, we can look in the local repository or folder and find that the image was stored there:

Figure 4.8: The image file we saved in the local repository or folder

This walk-through in Jupyter Notebook may seem like excessive steps, but taking the time to ensure the API is working and thinking through the logic of the common steps that need to be automated or repeated can be extremely beneficial when stepping into creating the Airflow DAG.

Let's step forward and automate this to run every day by creating an Airflow DAG.

Automating your code with a DAG

To turn this into a DAG, a few things need to be completed. These include setting up the DAG as originally designed as well as making use of some new tools that we have not covered yet around operators.

In most examples and code generation throughout this book, we will provide the code upfront and step through the information specific to the information provided throughout the chapter. The code for this can be found in the GitHub repository.

1. First, we begin by importing the required Python and Airflow libraries to build the DAG and create a variable called dag_owner to define who the author of the DAG is. We'll use the same libraries as our Jupyter Notebook exploration, as well as necessary Airflow imports:

```
import json
import pathlib
import airflow
```

```
import requests
import requests.exceptions as request_exceptions
from datetime import date
from airflow import DAG
from airflow.operators.bash import BashOperator
from airflow.operators.python import PythonOperator
from airflow.decorators import task
from datetime import datetime, timedelta


dag_owner = 'Kendrick'
```

2. Next, we define a Python function for getting the latest image from the NASA planetary API. Here, we call the API with our API key and save the image to a local file suffixed with today's date. We'll use this as the callable for the `PythonOperator` task in our DAG:

```
def _get_pictures():
    pathlib.Path("/tmp/images").mkdir(parents=True,
                                      exist_ok=True)
    api_key = 'xxxxxxxxxxxxxxxxxxxxxxxxxxxxxx'
    url = f'https://api.nasa.gov/planetary/apod?api_key={api_
key}'
    response = requests.get(url).json()
    today_image = response['hdurl']
    with open(f'todays_image_{date.today()}.png', 'wb') as f:
        f.write(requests.get(today_image).content)
```

3. Set the `default_args` for the DAG. This tells Airflow what parameters to use by default when calling each task in the DAG:

```
default_args = {'owner': dag_owner,
        'depends_on_past': False,
        'retries': 2,
        'retry_delay': timedelta(minutes=5)
        }
```

4. Initialize the DAG object parameters as a context manager using the `with` keyword. This prevents the Airflow scheduler from initializing DAG objects every time it parses the DAG code:

```
with DAG(dag_id='download_ASOD_image',
        default_args=default_args,
        description='download and notify ',
        start_date = airflow.utils.dates.days_ago(0),
        schedule_interval='@daily',
        catchup=False,
        tags=['None']
) :
```

5. Next, we can define the `get_pictures` task for the DAG using a `PythonOperator` and pass in our `get_pictures` function as the `python_callable`:

```
get_pictures = PythonOperator(
    task_id="get_pictures",
    python_callable=_get_pictures,
)
```

6. Then, we'll define a `notify` task to alert us that the image has been added. For simplicity, we just print a message with `BashOperator`, but this could easily be swapped with a different operator that alerts us another way, such as by email or Slack:

```
notify = BashOperator(
    task_id="notify",
    bash_command='echo f"Image for today has been
added!"',
)
```

7. Finally, we define the task execution path. This sets `get_pictures` as our first task and the `notify` task downstream from `get_pictures`:

```
get_pictures >> notify
```

Writing your first DAG

The Python code in the DAG may look much more intimidating than the Jupyter Notebook code that was originally provided, but most of the code is often repeated. Many of the items we are about to review are repeated fields that help Apache Airflow understand how it should function and it is up to you as a data engineer and data expert to determine the needs of your pipeline.

The best part about Airflow is that manual and high-volume tasks can be split up into multiple tasks that are automated. These tasks form together to create a DAG. In this example, the tasks are completed in sequential order, but in the more complex examples we will cover, Airflow is capable of completing the tasks in parallel.

Let's break down the DAG.

Instantiating a DAG object

The DAG is the beginning of any workflow pipeline and it must be identified so that Airflow knows which tasks belong to the DAG. You may see some DAGs starting with code similar to the following, but it is no longer recommended to use this method:

```
dag = DAG(
```

The new recommended method is to identify `with DAG(` and then identify the DAG name and associated DAG attributes. In this example, we label the DAG identification name, identify the default arguments that we will identify momentarily, and provide some other key information:

```
with DAG(dag_id='download_ASOD_image',
        default_args=default_args,
        description='download and notify ',
        start_date = airflow.utils.dates.days_ago(0),
        schedule_interval='@daily',
        catchup=False,
        tags=['None']
) :
```

A description is recommended to be added to DAGs as it provides an opportunity for you as a developer to provide information about what the DAG will accomplish on a frequent basis. It is a smart habit to provide a description of each set of code as well as comments to ensure other developers and team members can understand what the DAGs intend to accomplish.

The start date is set as the original date that the DAG should begin scheduling the tasks. In this example, it is set to 0 to allow for the DAG to start today. In other examples, it may be beneficial to explore starting multiple days ago. We will explore this idea more in further chapters.

Next, the schedule interval is set to identify the frequency the DAG will run. As a reminder, these can be set as preset values such as `@daily` or `@weekly` or as cron values. For this use case, we recommend using the daily function so it can take advantage of the daily updates. For other use cases, it may be more beneficial to run this only during the weekdays or only on weekends for other use cases, and Airflow can accomplish this uniqueness. We will get into these other details in further chapters.

`catchup` is an important field for Airflow and defaults to `False`. `catchup` defines whether Airflow should rerun the DAG for each interval in the past that is required. Since this example is set to start today, there is no need for `catchup`. In other cases, if we were to set the start date to be 7 days ago and the DAG to run daily, and `catchup` was set to `True`, we can expect the DAG to run 7 times upon the initial run.

Finally, the last DAG attribute identified is `tags`. Tags are a newer feature of Airflow and offer an opportunity for the data teams to better organize DAGs and provide tags for use. This can be extremely helpful for organizing DAGs by teams or by other functions.

Defining default arguments

In DAGs, some default arguments are separated and defined in a different area from the DAG instantiation. It is recommended to use this area to define other attributes of the DAG that are not as critical as the items identified previously but are important for the function of the DAG:

```
dag_owner = 'Kendrick'
```

```
default_args = {
        'owner': dag_owner,
        'depends_on_past': False,
        'retries': 2,
        'retry_delay': timedelta(minutes=5)
        }
```

In this example, the default arguments start with setting dag_owner to our co-author, Kendrick. This is an important field that should be kept as up-to-date as possible within your teams. Setting dag_owner allows your team to set different team members responsible for managing different DAGs and keeping them up to date. This is not always the same person as the original author of the DAG and can change frequently as teams change. It is recommended best practice to revisit the ownership of different DAGs every few months and understand who is responsible for DAG and task failure troubleshooting.

depends_on_past is a setting that, when set to True, keeps a task from becoming triggered if the previous schedule for the task has not succeeded. For this example, we want to set it to False so that if the DAG fails on one day, it still runs the next day. This is very beneficial for APIs that are not as dependable as others and ensures that the DAG still runs.

retries is a setting within Airflow to retry the task if it fails. In this example, we have chosen the value of 2. This means that if a task within the DAG fails, it will try two more times before exiting. In Airflow, the default value of retries is set to 0. This is another beneficial setting to enable if the API or connection is not reliable and is task-constrained.

Finally, retry_delay is set to 5 minutes in this example. This is to allow for 5 minutes to pass from the failure of a task before attempting to complete the task again. This is beneficial as the task can wait 5 minutes in between attempts in case the API is overloaded and working through other requests.

Defining the first task

The first task is set as get_pictures and completes almost all the steps that were identified in the Jupyter Notebook Python code from earlier. There are multiple ways to break up a set of code into tasks, but to keep this as simple as possible, we have elected to merge the actions into one task. Each piece of the Python code, from connecting, confirming connection, saving associated metadata, saving the image, and notifying, can be their own task. It is commonly recommended to break out the tasks as much as possible to help with troubleshooting:

```
get_pictures = PythonOperator(
    task_id="get_pictures",
    python_callable=_get_pictures,
)
```

As you will notice, the task name is set as the first value and set to a Python operator with multiple pieces of information within. `task_id` is set to `get_pictures` allowing us to label the task. Next, the Python callable is set to `_get_pictures`, which calls to the Python function, which we will cover in the next section after introducing operators.

What are operators?

Operators are among the fundamental building blocks of Airflow as they encapsulate the logic for performing specific tasks within a workflow, such as executing a Python function, running a Bash script, or sending an email. These operators are reusable and customizable, enabling complex workflows to be built with relative ease and limiting the amount of special and custom code creation.

Operators bring a breadth of value to Airflow. Operators abstract away the underlying complexity of task execution, allowing more focus on the high-level workflow logic. Operators exist to perform a single piece of work per task. In some cases, the operator may run custom Python code or a Bash script; in other instances, they can facilitate sending emails from your workflow.

Airflow has a large number of publicly available operators, handling various tasks. Among the most used operators are the following:

- `PythonOperator`: This versatile operator executes Python functions, allowing you to encapsulate complex logic and data processing within your workflow

- `BashOperator`: This operator executes Bash scripts, enabling you to leverage shell commands and scripts for various tasks, such as file manipulation, system administration, and tool execution

- `EmailOperator`: This operator facilitates sending emails from your workflow, allowing you to notify stakeholders, trigger alerts, or send reports

- `SimpleHttpOperator`: This operator makes HTTP requests to external APIs or web services, enabling you to integrate with third-party systems and retrieve data or perform actions

- `SnowflakeOperator`: This operator executes SQL queries against Snowflake databases, allowing you to perform data transformations, analyses, and manipulation within your workflow

- `KubernetesPodOperator`: This operator runs tasks within Kubernetes Pods, enabling you to leverage the scalability and resource management capabilities of Kubernetes for your workflow tasks.

- `MySqlOperator`: This operator executes SQL queries against MySQL databases, allowing you to perform data operations within your workflow

- `PostgresOperator`: This operator executes SQL queries against PostgreSQL databases, enabling you to perform data operations within your workflow

- `S3FileTransformOperator`: This operator transforms files stored in Amazon S3, enabling you to perform operations such as data cleaning, formatting, or conversion

- `SlackAPIOperator`: This operator interacts with the Slack API, enabling you to send notifications, post messages, or perform actions within your Slack workspace

Defining the first task's Python code

Within the first task, `get_picture`, we call the `_get_pictures` function with `PythonOperator`. Let's take a look at this code snippet:

```python
def _get_pictures():
    pathlib.Path("/tmp/images").mkdir(parents=True, exist_ok=True)
    api_key = 'xxxxxxxxxxxxxxxxxxxxxxxxxxxxxx'
    url = f'https://api.nasa.gov/planetary/apod?api_key={api_key}'
    response = requests.get(url).json()
    today_image = response['hdurl']
    with open(f'todays_image_{date.today()}.png', 'wb') as f:
        f.write(requests.get(today_image).content)
```

In this set of Python code, our goal is to connect to the API, locate the image, extract the image, and store it in a local folder to Airflow. It starts with ensuring a folder is created for storage. In this example, we take advantage of local storage, though in more complex examples, connections will have to be made to store the files in different storage locations.

Next, since this is a basic example, we set the API key within the function. In the next chapter, we will go through some examples of how to manage API keys and secrets in a more scalable manner. To utilize this code, we recommend replacing the `x` characters with your API key. The API key is used by the Python function to access the NASA API.

Next, we set the URL of the API and include the API key. For more information on using the APOD API, feel free to navigate to their GitHub site: `https://github.com/nasa/apod-api`.

With the JSON we reviewed previously in the Jupyter Notebook run of the Python code, we learned that `hdurl` is the location of the high-definition astronomy picture of the day, which meets our user needs. This URL is added to the `today_image` variable. Last, with the image opened, we write and store the image.

Defining the second task

With the image downloaded and stored as expected, we want to ensure we run a task to notify our users of the completion. This is a great opportunity to try out another operator:

```python
notify = BashOperator(
    task_id="notify",
    bash_command='echo f"Image for today have been added!"',
)
```

In the preceding code snippet, we set the second task as `notify` and utilize `BashOperator` to complete Bash console commands. With this operator, we are able to utilize `bash_command` to report that the image has been added.

To view the notification, we will navigate to the logs of Airflow. With the updated versions of Airflow, DAG and task logs can be reviewed within the DAG content screen, by selecting the **Grid** view, the task intended to be reviewed (in this case, the `notify` task), and selecting **Logs** from the navigation pane.

Inside of this set of logs, we are able to analyze the completion of the task along with our Bash command, which is shown about halfway down with `Images have been added!`.

Figure 4.9: DAG run logs for the notify task

Setting the task order

The last item to cover in this first DAG creation is to set the order in which the tasks are completed. In this example, we decided to explicitly state the order in which to be completed with the use of the `>>` icons:

```
get_pictures >> notify
```

This is the recommended practice for simpler and smaller sets of code. There are multiple other methods that are available, which we will cover in future chapters. To ensure this order is adhered to by Airflow, you can check the **Graph** view within Airflow. If all code was followed correctly and Airflow received your DAG code, you can expect a DAG similar to the following:

Figure 4.10: Simple DAG

Summary

In this chapter, we spent time working on creating our first DAG. We learned about creating an API key, how DAGs are constructed, and how operators work in relation to DAGs. When creating DAGs, it is important to remember that the way the tasks are broken up and created is up to the DAG author and maintainers of the code. It is much easier to troubleshoot multiple small tasks than large monolithic tasks. In addition, a consideration is to always work toward an architecture that meets both your and your user's needs from the pipeline.

In the next chapter, we will expand on the notification task and show ways to utilize notifications in your working teams with Slack and email notifications.

Part 3:
Common Use Cases

This part has the following chapters:

5

Connecting to External Sources

One of Apache Airflow's most powerful tools and functions is its ability to connect to external sources and hooks. This allows powerful workflow management for data engineers to connect to external sources such as databases, APIs, and cloud storage providers. It also provides the ability to call on other tools within the data engineering ecosystem such as data transforming tools, such as dbt, and **machine learning** (ML) tools, such as TensorFlow.

At its most basic level, connectors can be used to call upon secrets and perform notification functions for your team. In this chapter, we will look at the basics of setting up a connector within the Apache AI **user interface** (**UI**), authoring some additional lines of code to alert the team of a DAG failure, and reviewing best practices for storing secrets.

In this chapter, we are going to cover the following main topics:

- Adding connectors to the Apache Airflow UI
- Expanding on previous DAG examples
- Understanding the basics of secrets management within Airflow

Technical requirements

Similar to previous chapters, we expect that you have an Apache Airflow environment set up on your local machine and understand how to access it. If you have not done so already, please review *Chapter 5* as this code builds upon that example.

In this section, we will go into detail on how to set up connectors within Airflow. If you are looking for a more in-depth walk-through, then please navigate to the GitHub link: `https://github.com/ PacktPublishing/Apache-Airflow-Best-Practices/tree/main/chapter-05`.

Connectors make Apache Airflow

Being able to connect to external sources and tools is what makes Apache Airflow a versatile tool and orchestrator. It also reenforces the need to complete all compute outside of Apache Airflow. Some key examples of connections that Apache Airflow can make are as follows:

- Connecting to an external database and extracting data for a data warehouse, data lake/lakehouse, or data mart
- Connecting to an API similar to our example and retrieving data for an ML model
- Connecting to a service provider with webhooks such as Microsoft Teams or Slack

Computing outside of Airflow

Running compute-heavy tasks outside of Airflow is a key way to ensure the orchestrator meets your team's needs at scale. This is critical for performance as Airflow was designed to manage workflows, not to run heavy compute tasks. Running these compute tasks inside of Airflow can slow down workflows and make them prone to failure. In addition, Airflow costs resources and funding to run at scale, and these costs can quickly balloon if tasks are being run on compute within Airflow. Airflow at scale can be prone to failure if certain parameters are not set up to ensure your environment meets your team's needs.

We recommend running compute tasks outside of Apache Airflow. It is best to use a dedicated compute platform such as Spark or Hadoop or a cloud-based compute platform such as Google Cloud Dataproc or Amazon EMR. Save Apache Airflow for what it was developed to do: orchestrate and automate tasks.

Where are these connections?

Connections specific to Apache Airflow are objects that are used for storing credentials and other information that is needed for connecting to external services. Connections concerning the management of Airflow can be stored as environment variables, external secrets backend, and in the Airflow metadata database. Each of the three options has benefits and pitfalls. Let's take a look at storing within the Airflow metadata database first, which can be achieved through the UI or the **command-line interface** (**CLI**).

Connections stored in the metadata database

To find the connections section of Apache Airflow, navigate to the top bar of the Airflow UI. By selecting **Admin**, the drop-down menu will show **Connections**. Inside this area is where you can manually add connections to Airflow and review any current connections.

As a quick note before we continue, Airflow also allows connections to be configured via environment variables and with a secrets backend, such as AWS Secrets Manager. These methods are typically used in larger deployments of Airflow and will be referenced more as this book continues. Connections stored in environment variables or a secrets store will not appear in the connections list within the UI. These connections are still available to Airflow DAGs and tasks, but will not be visible to an admin within the UI console.

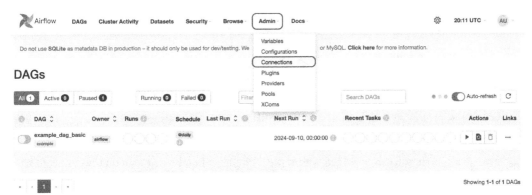

Figure 5.1: The Connections location for Apache Airflow UI

By selecting **Connections**, the user is taken to the complete list of connections. This information is available to any user within the organization who has access to the Airflow console. The connection passwords and sensitive variables are masked in this view. A user can choose to add, delete, or duplicate a connection. Duplicating connections may help with copying a template.

To create a new connection, select the + to add a new record.

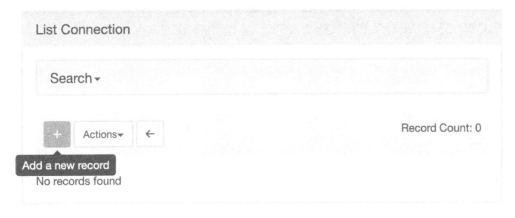

Figure 5.2: Add a new record through the Apache Airflow UI

Once selected, the user is prompted with a set of free text fields. Let's take a closer look at each one:

- **Connection Id**: A required free text entry field for the labeling of the connection. Many hooks have a default Connection ID that should be used, such as `slack_id` (used by Slack) and `postgres_default` (used by PostgresHook). This Connection ID is how you will later reference the connection from the DAG.

- **Connection Type**: A required drop-down menu selection. There are options for major cloud providers, such as AWS, GCP, and Azure, as well as connectors for HTTP and FTP.

- **Description**: Optional free text field for providing information about the connection.

Depending on the selection from the drop-down menu, the user will be prompted with more fields to enter information. For example, if the user has selected **Amazon Chime Webhook** as the connection type, they will be prompted with a request for the **Chime Webhook Endpoint**, **Schema** and **Chime Webhook token**.

Once created, a connection can be edited within the UI by selecting the edit icon on the connection. A connection can be deleted by selecting the delete icon as well.

A quick note about secrets being added through the Airflow UI

How your team goes about storing secrets is a critical function of how to approach any solution. In Apache Airflow, there are multiple places to store secrets, and we will dive deeper into this topic within this chapter.

In some connection types, a username and password are required to create and maintain the connection to a source. If the username and password are entered into the Airflow UI, the information will be saved within the Airflow metadata databases. All sensitive data within the Airflow metadata database is encrypted with a Fernet key, which can be rotated.

The following example is a screenshot of a new connection being created with a login and password being provided. The password is masked automatically by Airflow but is not encrypted when entered through the UI.

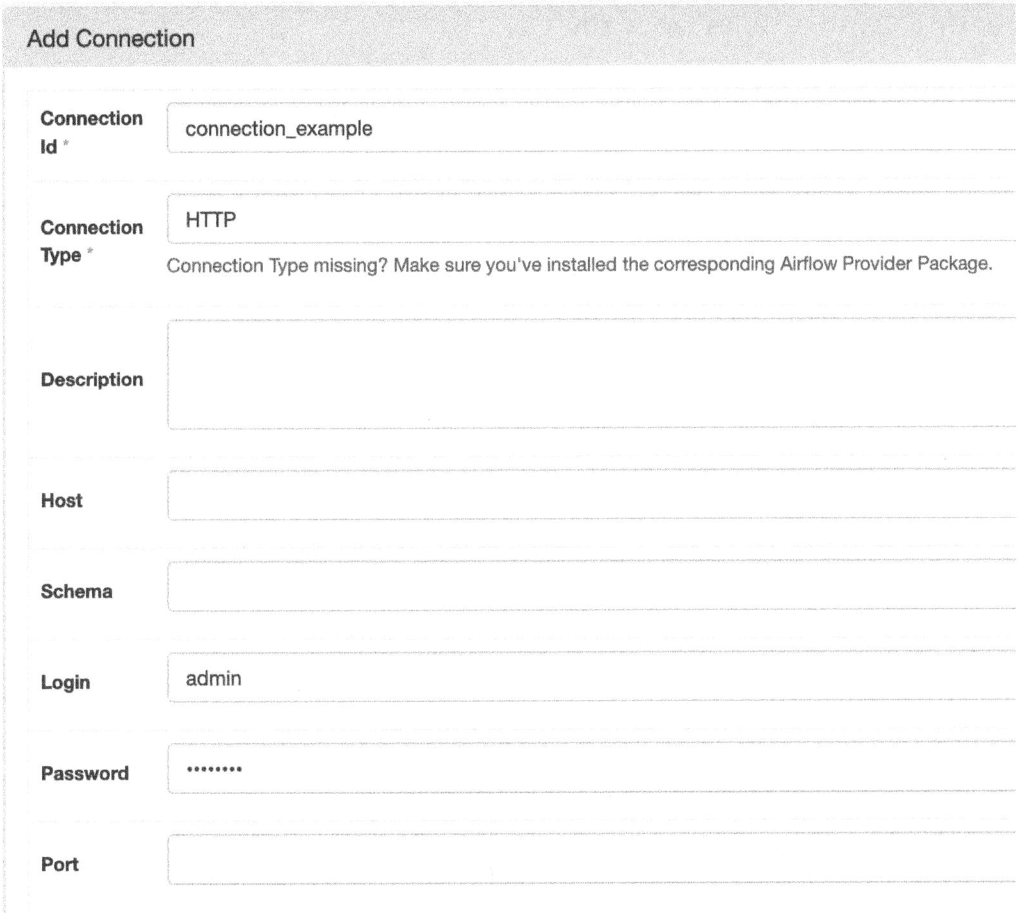

Figure 5.3: Add Connection with username and password

Once the Connection has been saved, a user can edit it. If a user wants to edit a connection that has a username and password, they will not be able to see the password within the connection settings. Airflow automatically masks Connection passwords, sensitive variables, and keys when they appear in task logs.

Creating Connections from the CLI

There have been a large number of improvements in the CLI for Apache Airflow over the past few years. Specifically, the focus on maintenance and changes from the CLI has been a focus and key area. Authenticated and connected users are able to add connections to the database directly from their CLI. Connections can be added using a JSON format, a URI format, or individually identifying each parameter.

An example of the JSON format, similar to our preceding connection, is as follows:

```
airflow connections add 'my_connection_db' \
    --conn-json '{
        "conn_type": "HTTP",
        "login": "admin",
        "password": "password",
        "host": "example-host",
        "port": 3306,
        "schema": "my-schema"
    }'
```

Learning how to use the CLI for Airflow Connections is a major benefit. Quick edits can be made to existing connections, new connections can be created through multiple formats, and you can export all connections from an airflow environment for migration. This is extremely helpful when migrating from an open source Airflow environment to a fully managed service. These connections can be exported as JSON, YAML, or env. To export, use the following:

```
airflow connections export connections.json
```

Alternatively, to export in a different file type, such as YAML, use the following:

```
airflow connections export /tmp/connections --file-format yaml
```

Testing of Connections

When creating, editing, or migrating connections, it is critical to test to ensure that the connection is successful. There is no worse feeling than receiving an alert for a failed task and realizing it's due to an incorrect parameter.

To test connections for your Airflow environment, there are a few ways to approach it. The most popular option is to utilize Postman when working to test an API connection. It allows the user to create, send, and receive API requests.

Another option is to test the connections within Airflow. This feature is disabled by default for the Airflow UI, API, and CLI as a security decision. To enable the feature, an update to the Airflow configuration file, `airflow.cfg`, is needed. Specifically, the `AIRFLOW__CORE__TEST_CONNECTION` field must be updated from **Disabled** to **Enabled**.

As an Airflow UI user, you can check the status of the flag by navigating to **Admin** and selecting **Configurations**. Within the list of configurations, the `test_connection` field will show the current setting.

core	test_connection	Disabled

Figure 5.4: The test_connection feature setting

Once enabled, connections can be tested within the Airflow UI or CLI. Within the UI, when creating or editing a connection, select the test connection button to test. Airflow calls the `test_connection` method from the selected hook class and will report the results of the test.

This tool is useful when it comes to early development and connections. From a secure environment perspective, it is not recommended to enable this feature. In addition, this feature is not useful for connections within external secrets backends.

As we have stepped through creating, editing, and testing connections within the Airflow UI and CLI, it is important to continue thinking about security. Securing secrets, such as usernames and passwords, or access keys is an important consideration when it comes to Connection management.

Please note that the test connection feature does *not work* for connections stored in environment variables or secret backends. These connections cannot be tested directly through the Airflow UI or CLI. As an alternative, you can create a dummy DAG that uses the connection in a task to verify that the connection is working. This approach is especially useful for connections stored in secret management systems.

Where we store secrets

Secrets are pieces of sensitive information that are used in the operation of Airflow and are often used as key information for your DAGs to successfully run. Secrets are variables, connections, and configurations for Airflow.

Apache Airflow offers three primary ways to store secrets: within the metadata database, as environment variables, or within an external secrets manager. Deciding which option is best for your team and situation is key as considerations need to be made around security, data at rest, network costs, and performance.

Using environment variables

One way to store secrets with Airflow is through environment variables, but it is the least secure. This method is only recommended for small, temporary test environments and getting started. Environment variables are set at the operating system level and can be accessed by any process running on the system.

To use environment variables to store secrets with Airflow, you can set them in the Airflow configuration file, `airflow.cfg`, or in the environment variable file, `.env`, which is called when Airflow is starting. These are not accessible from the Airflow UI or CLI and cannot be edited.

Using this method is quick and simple to set up, but it does not provide much security. In addition, if you are working in a large environment with multiple team members, this can create issues and key security risks for bad actors.

To store a connection or secret using environment variables in Airflow, the naming format needs to follow a specific convention. Connections should be named `AIRFLOW _CONN_<CONN_ID>` and variables should be formatted `AIRFLOW_VAR_<VARIABLE_NAME>` within the Airflow configuration file. This formatting allows for Airflow to automatically detect and use these connections and environment variables.

Another example is a connection string for a MySQL database, which would look like the following:

```
AIRFLOW_CONN_MYSQLDB = mysql://user:psasword@host:port/database_name
```

Following this format tells Airflow how to connect to the MySQL database by using the established protocol.

Airflow metadata database

As covered earlier in this chapter, the Airflow metadata database can be accessed from the Airflow CLI, UI, or API. Connections are easily added, edited, and deleted, whether in the UI or via the CLI. These are encrypted via the Fernet key (which is set during initial installations) in the database.

This option is good for teams looking to get started quickly but often fails to meet security standards and scale well. It is slower to set up than environment variables, but it is a good option to get started fast, and we will take a look at using this in our example.

Secrets management service

A **secrets management service** is a more secure option compared to using environment variables. Airflow supports multiple options, including AWS Secrets Manager, HashiCorp Vault, and Google Cloud Secrets Manager. Airflow was designed to orchestrate across multiple tools and cloud platforms; so, it is important that key secret information is stored appropriately, and access is limited.

This option is often the most secure and lives outside of the Airflow deployments, allowing for the data to persist even after an Airflow environment is torn down. In addition, this option provides the ability to rotate, version, and audit the certs and secrets.

Another significant advantage is that using a secrets management service allows you to keep connections in sync across multiple environments. For instance, if you store your development connections in a secret store and have every team member's local environment connected to it, you can easily add or update connections without needing to manually configure each user's machine. This ensures consistency and security across development, staging, and production environments.

To set up a secrets management service in Airflow, you need to do the following:

1. Set an environment variable in Airflow to point to the secrets store. This can be done by configuring the `AIRFLOW__SECRETS__BACKEND` environment variable to specify which secrets backend you are using.

2. Configure the secrets in the secrets store with the correct prefix so that Airflow can recognize them. For example, Airflow expects secrets to follow certain naming conventions, such as `airflow/connections/<CONN_ID>` for connections or `airflow/variables/<VARIABLE_NAME>` for variables.

Secrets Cache

As of writing this book, a new feature was recently released called **Secrets Cache**, introduced in Airflow 2.7. It is an experimental feature, disabled by default in the configuration file, with the goal of improving the performance of external secrets management and cost-savings.

Fetching secrets from an external secrets manager is a network operation, which can block the scheduler and create performance issues as this operation looks for the right return value. Caching these secrets at DAG parsing time can alleviate this issue and bottleneck. As teams scale and increase the total number of DAGs, the total parsing time increases as well as the costs associated. Fetching a secret can take over 100ms depending on the storage and requesting location.

How to test environment variables and secret store Connections

When using environment variables or secret stores to manage your Airflow connections and secrets, testing those connections slightly differs from testing connections stored directly in the Airflow metadata database. Since these connections do not appear in the Airflow UI or CLI, you cannot use the built-in **Test Connection** button for validation. Instead, you'll need to follow alternative approaches to ensure your connections are correctly configured and working as expected.

Testing environment variable connections

You can test environment variables storing connections by creating a **dummy DAG**. This will allow you to verify that Airflow can properly access and use the connection in an actual workflow. Here's how to do it:

1. **Create a simple DAG**: This DAG should use the connection stored in the environment variable. For example, if your environment variable is AIRFLOW_CONN_MYDB, create a task that utilizes this connection in a sample workflow.

2. **Trigger the DAG**: Trigger this in Airflow. This will run the task and attempt to use the connection.

3. **Check the task logs**: Do this to confirm the connection was established. If the connection is incorrect or misconfigured, the logs will indicate an error, such as a connection failure.

Here's an example of a simple DAG to test the connection:

```python
from airflow import DAG
from airflow.operators.python import PythonOperator
from datetime import datetime

def test_connection():
    import os
    conn_string = os.getenv("AIRFLOW_CONN_MYDB")
    print(f"Connection string: {conn_string}")

dag = DAG(
    'test_env_var_connection',
    start_date=datetime(2024, 1, 1),
    schedule_interval=None,
)

test_conn_task = PythonOperator(
    task_id='test_connection',
    python_callable=test_connection,
    dag=dag,
)
```

This DAG will print the connection string to the task logs, allowing you to confirm that Airflow is recognizing the environment variable.

Testing secret store Connections

Testing connections from a secrets store follows a similar process. However, since these connections are stored in an external system, the test will ensure that Airflow can retrieve and use the secret correctly.

Ensure that Airflow is correctly configured to use the secret store by setting the appropriate environment variables, such as AIRFLOW__SECRETS__BACKEND, and other necessary backend-specific settings (e.g., AWS region for AWS Secrets Manager).

Create a dummy DAG to use the connection or variable stored in the secrets management service. For instance, if your connection is stored in AWS Secrets Manager under the name airflow/connections/mydb, create a DAG task that tries to retrieve and use this connection.

Run the DAG and monitor the logs for the task. Check for any connection errors or missing secrets, which would indicate that Airflow is unable to retrieve the connection from the secrets store.

Here's an example DAG that tests a connection stored in a secret store:

```
from airflow import DAG
from airflow.operators.python import PythonOperator
from datetime import datetime
from airflow.hooks.base import BaseHook


def test_secret_connection():
    conn = BaseHook.get_connection("mydb")   # Replace "mydb" with the
actual connection ID from the secret store
    print(f"Connection Host: {conn.host}")

dag = DAG(
    'test_secret_store_connection',
    start_date=datetime(2024, 1, 1),
    schedule_interval=None,
)

test_secret_task = PythonOperator(
    task_id='test_secret_connection',
    python_callable=test_secret_connection,
    dag=dag,
)
```

This DAG retrieves the connection from the secret store and prints out the connection host, allowing you to confirm that Airflow is correctly accessing the secret.

Best practices

A best practice to keep in consideration when selecting the right secrets storing solutions is to consider your business needs and goals. If you are a large team looking to scale quickly and securely, then a secrets manager may be the best option. If you are a solo developer seeking to quickly test on a local machine, the metadata database or environment variables may be a better option.

With this review of best practices and secrets management, let's practice setting up a connection for a Slack alert.

Building an email or Slack alert

Teams and organizations setting up Apache Airflow are often beginning their journey without a wealth of experience or examples to lean upon. A poorly implemented and planned DAG can often lead to late nights of troubleshooting and frustrating returns to work realizing a task failed to complete due to incorrect setup. We will cover some best practices to remediate these issues in future chapters, but we often recommend setting up a Slack, Microsoft Teams, or email alert system to notify yourself or the team about failed task runs.

Walking through this example will help you set up a Slack alert and learn about setting up connections within the Airflow UI or CLI. If you require additional information, please navigate to the GitHub repository for a more thorough walkthrough.

Some key reasons why you may want to set up an alerting system such as this include the following:

- **Identify and respond to task failures quickly**: Alerting and notifications help to identify if an upstream task has failed or a network error is encountered, thus causing a problem with the overall system

- **Reduce system risk**: It helps to reduce the risk of insider threats, data loss, and other costly incidents

- **Comply with regulations**: Some industries require or have regulations requiring organizations to have alerting and notification systems in place

Key considerations

A key note to consider before diving into this example: Adding this example to DAGs will benefit your team by providing a key indication of when a task or DAG has failed to complete. This addition will not let you know if the Airflow environment has crashed or failed as it is an additional task.

Also, please be considerate to your fellow colleagues when implementing this feature. We have witnessed Slack channels and email alerts that provide no value to the team and often create more problems than solutions. Consider what this feature's goal is when implementing it and ensure that it meets your business needs.

Airflow notification types

Airflow can provide notifications on a task or DAG level. These include notifications for failures, retries, successes, on-execution, **service level agreements (SLAs)**, and timeouts.

One of the key types of notifications involves SLAs. An example of an SLA in Airflow is a time limit you can define for a task. Airflow will trigger a notification if a task takes longer than the specified SLA time to complete. This can notify you when something is running outside the expected performance window, ensuring that long-running tasks are identified and addressed promptly. SLA notifications can alert you to bottlenecks or issues in your workflow that could impact overall performance.

Notifications don't have to be sent out over Slack; other methods include email, Microsoft Teams, PagerDuty, and custom-made.

Email notification

Email notifications are built-in directly with Airflow functionality. To use this feature, you need to configure an SMTP server:

- As a demo, we will use a Gmail account. Start by creating a Google App Password for your Gmail account: `https://support.google.com/accounts/answer/185833`.

- Visit your App Password page for Gmail: `https://security.google.com/settings/ security/apppasswords`.

Under **Your App Passwords**, click **Select App** and choose **Mail**. Next, choose **Select Device** and choose **Other (Custom Name)**. You can name it anything but for this example, we will use `Airflow_ Email_Connection` and complete the app generation.

Copy and save the 16-character code that is generated for the app password and finish the process by selecting **Done**:

1. For this example, we will use an insecure form of secret storage that is not recommended at scale. Update your `airflow.cfg smtp` section to reflect the parameter update.

```
[smtp]
smtp_host = smtp.gmail.com
smtp_starttls = True
smtp_ssl = False
smtp_user = EMAIL_ADDRESS
smtp_password = YOUR_APP_PASSWORD
smtp_port = 587
smtp_mail_from = EMAIL_ADDRESS
```

Once saved and Airflow is updated, you will be able to take advantage of the default email notifications and alerts. These default notifications are defined in the `email_alert()` and `get_email_subject_content()` methods of the `TaskInstance` class. These can be overwritten by updating the `subject_template` and `html_content_template` variables in `airflow.cfg`.

2. To quickly implement a form of this example we set up, let's practice with a DummyOperator DAG:

```python
from datetime import datetime
from airflow import DAG
from airflow.operators.dummy import DummyOperator
from airflow.operators.python import PythonOperator
from airflow.operators.email import EmailOperator

def print_hello():
    return 'Hello world!'

default_args = {
    'owner': 'Kendrick',
    'start_date':datetime(2023,11,11),
}
with DAG(
    dag_id='email_alert_example',
    schedule_interval=None,
    default_args=default_args,
) as dag:

    email = EmailOperator(
        task_id='email_alert',
        to='smtp_user',
        subject='Email Alert',
        html_content="""<h3>Email Test</h3>""",
    )
    dummy_operator = DummyOperator(
        task_id='dummy_task',
        retries=3,
        dag=dag
    )
    hello_operator = PythonOperator(
        task_id='hello_task',
        python_callable=print_hello,
        dag=dag
    )
    email >> dummy_operator >> hello_operator
```

At the end of this section, you'll now have email configured and a test case for ensuring that your email notifications are being delivered. This is a critical first step in operationalizing Airflow for production workloads.

Creating a Slack webhook

To begin, let's update our `requirements.txt` file to include the HTTP and Slack providers in our airflow environment. Amend the file with the following lines:

```
apache-airflow-providers-http
apache-airflow-providers-slack
```

After making these changes, the Airflow environment must be restarted.

Next, we will want to create a Slack webhook to send data to. Follow these instructions to register and create your Slack app: `https://api.slack.com/docs/slack-button#register_your_slack_app`.

Once complete, create an incoming webhook by clicking on **Add New Webhook to Workspace**.

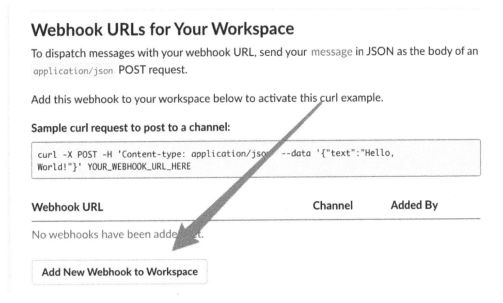

Figure 5.5: Add New Webhook to Workspace

Next, select the Slack channel you want to post alerts to. Be aware of who uses this channel and what value these alerts will provide. Once selected, Slack will provide you with a webhook URL. Copy this webhook URL and save it for the Airflow Connection creation. It will look similar to the following:

```
https://hooks.slack.com/services/T00000000/B00000000/
XXXXXXXXXXXXXXXXXXXXXXXX
```

Creating the Airflow Connection

With the Slack webhook created, we now need to go into the Airflow UI or CLI to create the connection. To complete this action, navigate to your Airflow instance, select **Admin**, and then select **Connections**.

```
Set the following parameters:
Connection Id = slack
Connection Type = HTTP
Password = T00000000/B00000000/XXXXXXXXXXXXXXXXXXXXXXXX
```

Add Connection	
Connection Id *	slack
Connection Type *	HTTP x ▾
	Connection Type missing? Make sure you've installed the corresponding Airflow Provider Package.
Description	
Host	
Schema	
Login	
Password	••
Port	

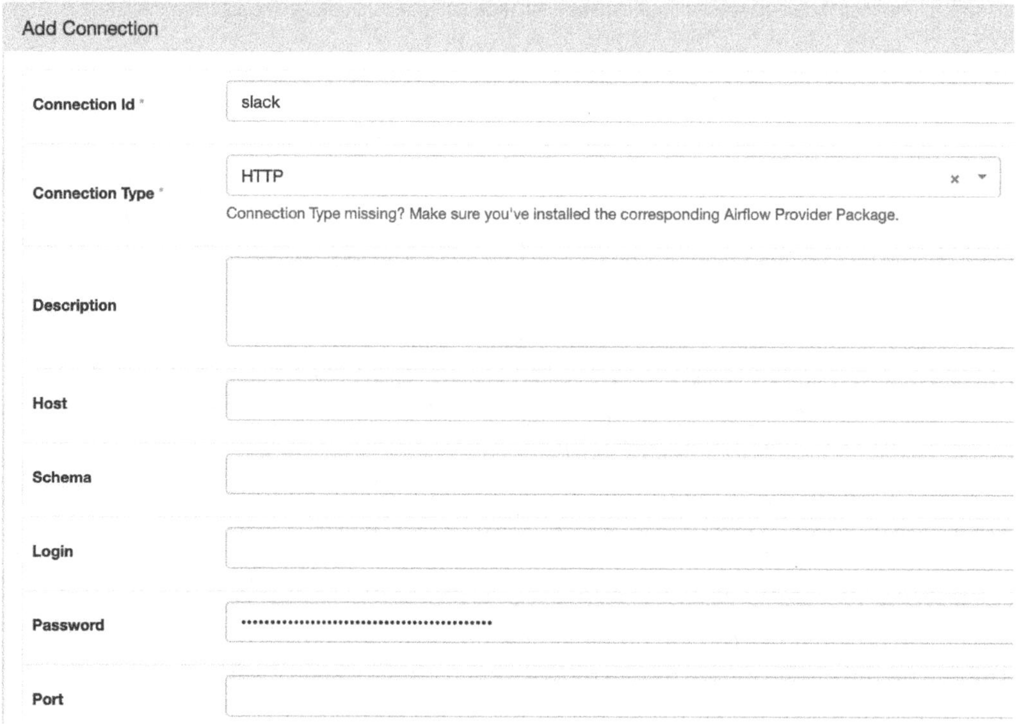

Figure 5.6: Image of an Airflow Slack Connection

If you prefer to create this connection via the Airflow CLI, the following JSON can be used:

```
An example of the JSON format, similar to our preceding connection, is
as follows:
airflow connections add 'slack' \
    --conn-json '{
        "conn_type": "HTTP",
        "password": " T00000000/B00000000/XXXXXXXXXXXXXXXXXXXXXXXX "
    }'
```

As a reminder, testing a connection is recommended. We recommend using an external test such as Postman or within the Airflow UI or CLI.

Let's build an example DAG

Let's put together a simple call-back function for this Slack alerting example. In the following excerpt, the Python code checks for and sends an alert to Slack for successful task completion.

As a callout, this DAG was originally authored by our colleague, Adam Spain. To review the complete GitHub, you can reference this link: https://github.com/aspain/airflow-slack-callbacks/blob/main/include/old_version/slack_callback_functions.py#:

```
from airflow.providers.slack.operators.slack_webhook import
SlackWebhookOperator
from airflow.hooks.base import BaseHook
def success_callback(context):
    slack_conn_id = 'slack'
    slack_webhook_token = BaseHook.get_connection(slack_conn_id).
password
    log_url = context.get('task_instance').log_url

    slack_msg = f"""
            :white_check_mark: Task has succeeded.
        *Task*: {context.get('task_instance').task_id}
        *Dag*: {context.get('task_instance').dag_id}
        *Execution Time*: {context.get('execution_date')}
        <{log_url}| *Log URL*>
        """

    slack_alert = SlackWebhookOperator(
        task_id='slack_test',
        http_conn_id='slack_callbacks',
        webhook_token=slack_webhook_token,
        message=slack_msg,
        username='airflow')
    return slack_alert.execute(context=context)
```

Once implemented and registered as the successful callback parameter. When the dag is successfully run, an alert similar to the following will appear in your Slack channel for each successful execution.

Figure 5.7: An example of the Slack message that will appear when your DAG has succeeded

Summary

In this chapter, we covered some key areas of the Airflow architecture and the primary components that make up connections and secrets. We spent time analyzing what Connections are, how to create, edit, and delete a connection, and how they are best used. In addition, secrets are often needed to create and maintain connections. There are multiple ways and tools available to meet your team's needs and business goals, but the following chart may help with future decisions.

Method	Pros	Cons
Environment Variables	Simple and easy to set up and use	Not very secure, with no access from Airflow UI or CLI
Metadata Database	Quick to set up and edit values via the UI and CLI	Not secure and hard to scale
Secrets Management Service	Secure, hosted outside of Airflow, and easy to keep in sync	Higher costs and no access from the Airflow CLI or UI

Finally, we stepped through two examples of email and Slack alerting. Each example is powerful in alerting organizations of issues with tasks and DAGs but may have shortfalls in the way these alerts are set up and implemented. We recommend always revisiting your alert management, even on a monthly or quarterly cadence to assess how helpful they are.

In the next chapter, we will step into a more robust ETL pipeline and learn how connections support them.

6

Extending Functionality with UI Plugins

Apache Airflow provides a robust and extensible web-based **user interface** (**UI**) that allows users to monitor and manage their workflows. While the default Airflow UI offers a wide range of essential features, there are times when you need a more specialized view to gain insights into the performance and health of your workflows.

In this chapter, we will explore the process of creating a custom Airflow UI plugin to display a metrics dashboard. You will gain a basic understanding of HTML/CSS and JavaScript concepts and experience building custom web views with Flask blueprints.

This chapter covers the following topics:

- Understanding Airflow UI plugins
- Creating a metrics dashboard plugin

Technical requirements

To begin developing your custom plugin, you'll need a Python environment and a local Apache Airflow environment for testing. Please review *Chapter 5* to learn how to set up your local Apache Airflow environment.

The code for this chapter is available at `https://github.com/PacktPublishing/Apache-Airflow-Best-Practices/tree/main/chapter-06`. You should refer to the code as you read through this chapter to ensure you fully understand the implementations outlined.

Understanding Airflow UI plugins

Airflow UI plugins extend the capabilities of the default web interface, enabling users to add custom views and functionalities tailored to specific business requirements. Extending the Airflow UI is particularly useful when incorporating domain-specific metrics, monitoring tools, or visualizations into your Airflow environment.

Plugins need to be registered into a Python class that inherits from the `airflow.plugins_manager.AirflowPlugin` interface. The plugin class allows you to register blueprints, appbuilder views, custom hooks, and timetables and implement several other customizations that are described in the Airflow documentation. The following code block is an example of the `AirflowPlugin` class:

```
class AirflowPlugin:
    # The name of your plugin (str)
    name = None
    # A list of class(es) derived from BaseHook
    hooks = []
    # A list of references to inject into the macros namespace
    macros = []
    # A list of Blueprint object created from flask.Blueprint. For use
with the flask_appbuilder based GUI
    flask_blueprints = []
    # A list of dictionaries containing FlaskAppBuilder BaseView
object and some metadata. See example below
    appbuilder_views = []
    # A list of dictionaries containing kwargs for FlaskAppBuilder
add_link. See example below
    appbuilder_menu_items = []

    # A callback to perform actions when airflow starts and the plugin
is loaded.
    # NOTE: Ensure your plugin has *args, and **kwargs in the method
definition
    #    to protect against extra parameters injected into the on_
load(...)
    #    function in future changes
    def on_load(*args, **kwargs):
        # ... perform Plugin boot actions
        pass

    # A list of global operator extra links that can redirect users to
    # external systems. These extra links will be available on the
    # task page in the form of buttons.
    #
    # Note: the global operator extra link can be overridden at each
    # operator level.
    global_operator_extra_links = []

    # A list of operator extra links to override or add operator links
```

```
    # to existing Airflow Operators.
    # These extra links will be available on the task page in form of
    # buttons.
    operator_extra_links = []

    # A list of timetable classes to register so they can be used in
DAGs.
    timetables = []

    # A list of Listeners that plugin provides. Listeners can register
to
    # listen to particular events that happen in Airflow, like
    # TaskInstance state changes. Listeners are python modules.
    listeners = []
```

Airflow uses Python Flask to render the webserver UI. Flask applications utilize blueprints to render web elements, which allows us to group web app functionality into reusable components. Blueprints are assigned to API routes when registered with the Flask application and have resources such as static files, templates, and Appbuilder views. Appbuilder views are used to create the API routes for the Flask blueprints. By the end of this chapter, you'll understand how to link these elements together to create Airflow UI plugins.

Here is a list of use cases for creating custom Airflow UI plugins to enhance the features of Airflow:

- A Hive Metastore browser that allows users to view important details about Metastore tables or parse logs

- A monitoring integration page that includes buttons to create and manage monitors for DAGs in Datadog

- Visualizations of SLA statuses for all DAGs for an overall view of missed SLAs

Creating a metrics dashboard plugin

We're going to create a simple metrics dashboard that is accessible via the menu bar in our local Airflow environment. To easily generate the charts for the dashboard, we'll be using Chart.js. Chart.js is a simple and flexible charting library written in JavaScript. It provides a set of frequently used chart types, plugins, and customization options and a set of built-in chart types. More information about Chart.js can be found on the documentation page at `https://www.chartjs.org/docs/latest/`.

Let's outline the steps to create a custom Airflow UI plugin for a metrics dashboard.

Step 1 – project structure

Airflow plugins are stored in a `plugins` directory, which must be added to the Airflow home directory before Airflow starts up. This is because plugins are lazily loaded by default, and any changes to the plugin will require a restart of the Airflow services. Let's organize the `plugins` directory with the following structure:

```
plugins
└── metrics_plugin
    ├── __init__.py
    ├── templates
    |   └── dashboard.html
    └── views
        ├── __init__.py
        └── dashboard.py
```

Here, we have a directory for the plugin module called `metrics_plugin`. We will register the plugin and Flask blueprint in the `__init__.py` file. The `templates` directory contains the frontend HTML code for displaying the dashboard. We also have `views/dashboard.py`, where we will implement the backend code for the web view.

Step 2 – view implementation

Let's begin by implementing the backend web view in `views/dashboard.py`. The Flask application will execute the backend code when the registered route is visited in the webserver UI. This allows us to execute queries against the Airflow database for pulling the metrics displayed in the dashboard:

1. First, we will set up the imports that are required for the dashboard view:

    ```python
    from __future__ import annotations
    from typing import TYPE_CHECKING
    from airflow.auth.managers.models.resource_details import
    AccessView
    from airflow.utils.session import NEW_SESSION, provide_session
    from airflow.www.auth import has_access_view
    from flask_appbuilder import BaseView, expose
    from sqlalchemy import text
    if TYPE_CHECKING:
        from sqlalchemy.orm import Session
    ```

    The `has_access_view` decorator is used to confirm that the user has permission to access the website so that we don't expose the page without an authenticated user. The `provide_session` decorator will supply a database session argument to the route function to be used for queries. `BaseView` is the class that `MetricsDashboardView` will be derived from.

2. Let's set up the `MetricsDashboardView` class to define the route for the web view:

```
class MetricsDashboardView(BaseView):
    """A Flask-AppBuilder View for a metrics dashboard"""
    default_view = "index"
    route_base = "/metrics_dashboard"
```

Here, we define a class called `MetricsDashboardView` and set some default values for `default_view` and `route_base`. The `default_view` instructs Flask which function to use for the base route; in this case, we will define a function called `index`. To make the base route more user-friendly, `route_base` is set to `/metrics_dashboard`.

3. Finally, the index function will execute the queries we need to run against the Airflow database to provide our metrics:

```
@provide_session
@expose("/")
@has_access_view(AccessView.PLUGINS)
def index(self, session: Session = NEW_SESSION):
    """Create dashboard view"""
    def interval(n: int):
        return f"now() - interval '{n} days'"
    dag_run_query = text(
        f"""
        SELECT
            dr.dag_id,
            SUM(CASE WHEN dr.state = 'success' AND dr.start_date
> {interval(1)} THEN 1 ELSE 0 END) AS "1_day_success",
            SUM(CASE WHEN dr.state = 'failed' AND dr.start_date
> {interval(1)} THEN 1 ELSE 0 END) AS "1_day_failed",
            SUM(CASE WHEN dr.state = 'success' AND dr.start_date
> {interval(7)} THEN 1 ELSE 0 END) AS "7_days_success",
            SUM(CASE WHEN dr.state = 'failed' AND dr.start_date
> {interval(7)} THEN 1 ELSE 0 END) AS "7_days_failed",
            SUM(CASE WHEN dr.state = 'success' AND dr.start_date
> {interval(30)} THEN 1 ELSE 0 END) AS "30_days_success",
            SUM(CASE WHEN dr.state = 'failed' AND dr.start_date
> {interval(30)} THEN 1 ELSE 0 END) AS "30_days_failed"
        FROM dag_run AS dr
        JOIN dag AS d ON dr.dag_id = d.dag_id
        WHERE d.is_paused != true
        GROUP BY dr.dag_id
        """
    )
    dag_run_stats = [dict(result) for result in session.
execute(dag_run_query)]
    return self.render_template(
```

```
        "dashboard.html",
        title="Metrics Dashboard",
        dag_run_stats=dag_run_stats,
    )
```

dag_run_query pulls all DAG runs for active DAGs and does some aggregations for counting successful and failed executions for three separate time periods: 1 day, 7 days, and 30 days. We then convert the results of the query to a list of dictionaries called dag_run_stats and return the rendered template in the HTML described in the next step.

> **Important note**
> The dag_run_stats list is passed as an argument to the render_template function, which will allow the HTML to utilize the data from our query in step 2.3 within the Jinja template.

Step 3 – metrics dashboard HTML template

Now, we will create the HTML template for the metrics dashboard in templates/dashboard. html. Airflow utilizes Bootstrap CSS, which simplifies styling to support a responsive web experience and Jinja templating for rendering web views.

We need to add the Airflow base templates to the HTML file, which include the menu bar and other HTML head elements:

```
{% extends base_template %}
{% block title %}
{{ title }}
{% endblock %}
{% block head_meta %}
{{ super() }}
{% endblock %}
```

Now we need to define the content block where the charts will be displayed. The HTML required for the content block is very simple since we will be rendering the charts with JavaScript:

```
{% block content %}
<h2>{{ title }}</h2>
<div class="container-fluid">
  <div class="row">
    <div class="col-lg-6 col-md-12">
      <canvas id="successChart"></canvas>
    </div>
    <div class="col-lg-6 col-md-12">
      <canvas id="failedChart"></canvas>
    </div>
```

```
    </div>
  </div>
{% endblock %}
```

The class directives are from Bootstrap CSS to make the graphs responsive to different screen sizes. There are two canvas elements with the IDs successChart and failedChart where the successful and failed DAG runs will be rendered, respectively.

Next, we will add the tail block, where we will define the JavaScript required to render the charts:

```
{% block tail %}
{{ super() }}
<script src="https://cdn.jsdelivr.net/npm/chart.js"></script>
<script>
  const data = {{ dag_run_stats | tojson }};
  new Chart(
    document.getElementById('successChart'),
    {
      type: 'bar',
      title: "Successful Dag Runs",
      data: {
        labels: data.map(row => row.dag_id),
        datasets: [
          {
            label: "1 day success",
            data: data.map(row => row["1_day_success"])
          },
          {
            label: "7 days success",
            data: data.map(row => row["7_days_success"])
          },
          {
            label: "30 days success",
            data: data.map(row => row["30_days_success"])
          }
        ]
      },
      options: {
        responsive: true,
        indexAxis: 'y',
        scales: {
          x: {
            type: 'logarithmic',
            display: true,
            title: {
```

```
                    display: true,
                    text: "Number of Dag Runs"
                }
            }
        },
        plugins: {
          title: {
            display: true,
            text: "Successful Dag Runs"
          }
        }
      }
    }
  }
);
new Chart(
  document.getElementById('failedChart'),
  {
    type: 'bar',
    data: {
      labels: data.map(row => row.dag_id),
      datasets: [
        {
          label: "1 day failed",
          data: data.map(row => row["1_day_failed"])
        },
        {
          label: "7 days failed",
          data: data.map(row => row["7_days_failed"])
        },
        {
          label: "30 days failed",
          data: data.map(row => row["30_days_failed"])
        }
      ]
    },
    options: {
      responsive: true,
      indexAxis: "y",
      scales: {
        x: {
          type: "logarithmic",
          display: true,
          title: {
            display: true,
            text: "Number of Dag Runs"
```

```
                }
              }
            },
          plugins: {
            title: {
              display: true,
              text: "Failed Dag Runs"
            }
          }
        }
      }
    );
  </script>
  {% endblock %}
```

Here, we inject the `chart.js` library from the open source CDN so that we can use the library to render the charts in the canvas elements. We also define a data variable with the `dag_run_stats` JSON that we created in the dashboard view backend code. The last important piece is to call the new `Chart()` methods, which take the canvas elements as the first argument and the configuration of the chart as the second.

Step 4 – plugin implementation

The final step in building our metrics dashboard plugin is to register the Flask `Blueprint` and `MetricsDashboardView` with Airflow. Enter the following code in the `metrics_plugin/__init__.py` file:

```python
from __future__ import annotations

from airflow.plugins_manager import AirflowPlugin
from flask import Blueprint

from plugins.metrics_plugin.views.dashboard import
MetricsDashboardView

# Creating a flask blueprint
metrics_blueprint = Blueprint(
    "Metrics",
    __name__,
    template_folder="templates",
    static_folder="static",
    static_url_path="/static",
)
```

```
class MetricsPlugin(AirflowPlugin):
    """Defining the plugin class"""


    name = "Metrics Dashboard Plugin"
    flask_blueprints = [metrics_blueprint]
    appbuilder_views = [{
        "name": "Dashboard", "category": "Metrics",
        "view": MetricsDashboardView()
    }]
```

The preceding code contains a simple Flask `Blueprint` and a class that is derived from the `airflow.plugins_manager.AirflowPlugin` class. We register `metrics_blueprint` and `MetricsDashboardView` by including them in the `flask_blueprints` and `appbuilder_views` class variables, respectively. In the `MetricsPlugin` class, we've given the dashboard a name and a category, which will ultimately organize the Airflow menu item as **Metrics -> Dashboard**.

To integrate the plugin with Airflow, we need to place the `metrics_plugin` folder inside the Airflow home plugins directory. Then, we can restart the webserver and test that the page is working as expected. You'll see the **Metrics** category in the menu bar with the **Dashboard** menu item representing the metrics dashboard view.

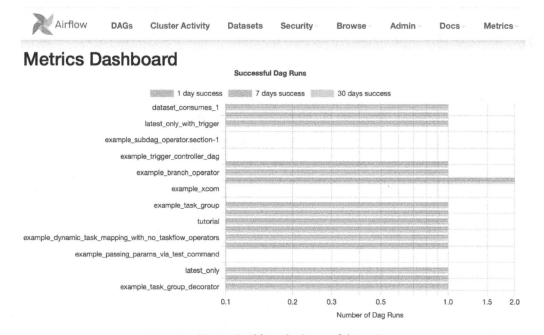

Figure 6.1: Metrics Dashboard – Successful Dag Runs

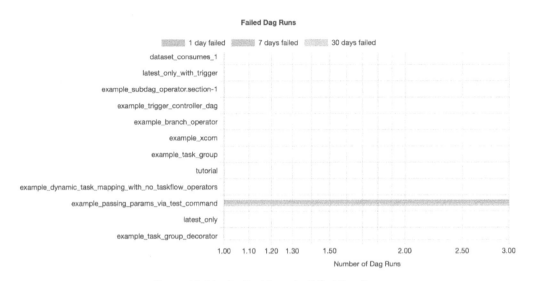

Figure 6.2: Metrics Dashboard – Failed Dag Runs

Our metrics dashboard is finally complete! In *Figure 6.1*, we can see the **Successful Dag Runs** graph depicting THE number of pipeline executions that have completed successfully over the last 1 day, 7 days, and 30 days. *Figure 6.2* shows the DAG runs that have failed over the same time frames (1 day, 7 days, and 30 days). To create the metrics dashboard, we executed the following steps:

1. **Project structure**: We set up the project structure in the Airflow plugins directory, following a pattern that can be reused to implement other custom plugins for our Airflow environment.

2. **View implementation**: We created `MetricsDashboardView`, which defines the backend functionality of our dashboard page and queries the database to retrieve the data for the graphs.

3. **Metrics dashboard HTML template**: We created the HTML template that builds the web view of the dashboard and contains the JavaScript to load the graphs with the data from the view query.

4. **Plugin implementation**: We registered the plugin in the `MetricsPlugin` class, which implements the `AirflowPlugin` class to add the web view to our Airflow environment.

Summary

By following the steps outlined in this chapter, you can build a powerful and visually appealing dashboard that provides valuable insights into the performance of your Airflow environment. Creating a custom metrics dashboard with Apache Airflow UI plugins allows you to tailor your monitoring experience to the unique requirements of your workflows. While the default Airflow UI provides essential features, the ability to create custom views tailored to specific business requirements is invaluable. We've created a simple dashboard that highlights DAG run success and failure over three separate time frames. Experiment with different visualizations and metrics to make the most of this extensible feature and enhance your workflow management experience.

In the next chapter, we'll explore how to develop and distribute custom provider packages so that other Airflow users can implement them in their Airflow environments. Building provider packages that can be installed via PyPI is also a great way to gain experience in contributing to the open source community.

References

1. Apache Airflow. (n.d.). Plugins — Apache Airflow Documentation. Retrieved January 28, 2024, from `https://airflow.apache.org/docs/apache-airflow/stable/authoring-and-scheduling/plugins.html#interface`

2. Chart.js. (2023, December 4). Retrieved January 28, 2024, from `https://www.chartjs.org/docs/latest/`

7

Writing and Distributing Custom Providers

As you continue your journey with Airflow, you will no doubt encounter times when the providers that are supported by the core airflow community do not suit your needs. Initially, you might extend functionality with a single function, in a localized Python module, which you execute via the `PythonOperator`. You might even go so far as to extend a base operator class within that same module and use it directly to instantiate tasks.

After a while, someone in another group at your company or maybe just a fellow developer on the internet may describe a problem where your solution is the perfect fit. The question is, *How do you get them that code in the best way possible?*.

In this chapter, we're going to show you how to build and distribute Airflow providers so that others can use it within their Airflow deployments. We're going to use an intentionally silly problem so that our solution can stay simple.

This chapter covers the following topics:

- Structuring your provider
- Authoring your provider

Technical requirements

This chapter will delve (lightly) into Python testing frameworks, as well as Python packaging and distribution. We will do our best to ensure that we explain salient points as we go, but it may be useful to do additional reading on these subjects.

Our local development environment also makes extensive use of Docker and Docker Compose. Functional knowledge of these tools may be useful.

Code for this chapter is available at `https://github.com/PacktPublishing/Apache-Airflow-Best-Practices/tree/main/chapter-07`. You should refer to it in its entirety as you read this chapter to gain a full understanding of what a provider entails.

Structuring your provider

Airflow providers are simply Python packages, with a few special interfaces and implementations to ensure that they integrate well into the core Airflow code base. If you've ever written a Python package for distribution, most of the steps and tasks within this chapter should feel familiar to you.

General directory structure

> **Important note**
>
> There are a lot of boilerplate directories, files, and code required to make a distributable package. As such, a number of templating engines (and templates) have been created to lower the burden of setting up projects. We use Angreal and that equips Airflow provider template to create this boilerplate. You can create your own provider with this system using the following commands:
>
> ```
> pip install angreal
>
> angreal init airflow_provider
> ```

Let's begin by looking at the important directories and files within our teapot provider package:

```
.
├── airflow_provider_tea_pot
│   ├── hooks
│   ├── __init__.py
│   ├── operators
│   ├── provider.py
│   ├── sensors
│   └── triggers
├── dev
│   ├── docker-compose.yaml
│   └── Dockerfile
├── example_dags
├── README.md
├── setup.cfg
├── setup.py
└── tests
```

Package folder

The package folder is the directory that contains the source code for your actual provider. In this case, it's `airflow_provider_tea_pot`. By convention, your provider should always start with `airflow_provider`, followed by the technology you're integrating with.

This folder may contain the following folders as modules:

- **Hooks**: For writing all of the hooks for the service you're integrating with
- **Operators**: For writing all of the operators that will be used in DAGs
- **Sensors**: For any specialized operators that wait on or instrument an external service
- **Triggers**: For triggers that will be used by specialized deferrable operators

In this example, we also include a single file module, `provider`, that includes a function that will be used to register your provider package with Airflow when it is installed.

Dev folder

In this example, the `dev` folder contains files and folders that are being used to provide a localized testing and demo environment for your provider. We provide a basic Dockerfile and `docker-compose.yaml` to describe this environment.

Example DAGs folder

This folder contains described DAGs that can be run within your development environment to demonstrate your functioning provider. These can be incredibly useful to help users get a concrete example of how to consume your provider and bootstrap their own experimentation.

Tests folder

This folder houses any sort of unit, functional, or end-to-end tests you write as part of your provider. In this example, we are using common idioms from the `py-test` framework for writing and executing any test cases. We strongly recommend that you write tests for any providers you author; it's a great way to demonstrate to yourself and others that your code functions as intended.

Packaging files

Two files are primarily responsible for making your code able to be packaged (and therefore distributed) in this repo: `setup.py` and `setup.cfg`. The `setup.cfg` file contains metadata about your package so that Python's build and distribution tools understand how to package and install it. The `setup.py` is not strictly necessary; it is useful when you wish to install your package in an *editable* mode, so we have included it here.

> **Important note**
>
> While technically, the naming conventions and submodule structures presented in this chapter are not entirely required for your provider to function within Airflow, these conventions have existed within the community for long enough to be considered standards. It is strongly recommended to follow them so that other Airflow users have an easier time understanding the code you've written.

So far, nothing about the structure of this Python package is particularly noteworthy; this is intentional. The Airflow community has been deliberate and thoughtful in designing and implementing its plugin system so that it uses completely standard Python idioms and practices to function. The following sections will delve into specific interfaces that are required to ensure that your provider fully integrates into your Airflow instance and functions well.

Authoring your provider

Now that we've got our project layout created, it's time to start modifying our packaging files and code to ensure that our provider package integrates well with Airflow.

> **General design considerations**
>
> Due to the way Airflow interacts with the code it executes, there are a few important things to keep in mind when authoring your provider.
>
> **Code should be able to run without access to the internet**. The Airflow Scheduler parses DAGs on a regular cadence. When that parsing occurs, anything in the `__init__` method of a class will be executed, including calls to third-party APIs and services.
>
> **Initialization methods should not call functions that only return valid objects during task execution**. If this is violated, you will encounter import errors when attempting to use a provider in a DAG. You should instead use Jinja templating and macros within your DAG to manage these kinds of configurations.
>
> **All operators require an** `execute` **method**. This method is a required interface for all operators.

Registering our provider

Before we write any code, we want to make sure that our provider registers itself with Airflow on startup. Airflow's provider specification expects this to happen through package metadata, specifically through a specific package entry point called `apache_airflow_provider`.

In your provider module (`provider.py`), create a single function called `get_provider_info` and have it return a dictionary with the following keys: `package-name`, `name`, `description`, and `versions`. These keys will provide Airflow with information to display from the UI, Airflow CLI, and API interface:

```
def get_provider_info():
    return {
```

```
       "package-name": "airflow-provider-tea-pot",
       "name": "Teapot Provider",
       "description": "`A short and stout provider for Pakt
Publication <https://github.com/PacktPublishing/Apache-Airflow-Best-
Practices>`__",
          "versions": airflow_provider_tea_pot.__version__,
   }
```

We also need to tell Python that this entry point exists. We can do this by adding the following information to our `setup.cfg` file:

```
[options.entry_points]
apache_airflow_provider=
provider_info=airflow_provider_tea_pot.provider:get_provider_info
```

Now that we've completed these steps, our package will register itself as an Airflow provider when present during Airflow starting up. This is the primary way that you will provide functionality for your provider within the Airflow UI, such as creating connections and providing links to external documentation.

Authoring our hook

Our smart teapot has a described API that can be interacted with via simple HTTP requests and appears to implement a minimal subset of the **Hyper Text Coffee Pot Control Protocol (HTCPCP)**:

Method	Endpoint	Description	Return
GET	/ready	Check whether your teapot is ready to brew	200 `{ "ready" }`
POST	/make_tea	Instruct your teapot to brew you a cup of tea	200 `{ "success" }`
POST	/brew_coffee	Instruct your teapot to brew a pot of coffee	418 `{ "I am a teapot." }`
GET	/water_level	Check the water level of your teapot	200 `{ float(0,1) }`

Table 7.1: A list of HTTP methods for our teapot service

With these exposed methods in mind, we can start by writing a hook that allows us to describe a connection to a specific coffee pot and send and receive specific messages to that pot.

We start by subclassing an Airflow base hook describing an initialization method to store information about the initialized class:

```
class TeaPotHook(BaseHook):
    conn_name_attr = "tea_pot_conn_id"
    default_conn_name = "tea_pot_default"
    conn_type = "teapot"
    hook_name = "TeaPot"

    def __init__(
        self,
        tea_pot_conn_id: str = default_conn_name,
    ) -> None:
        super().__init__()
        self.tea_pot_conn_id = tea_pot_conn_id
```

It is a fairly standard practice to set hook_name, conn_type, default_conn_name, and conn_name_attr to the class scope, especially if you plan on further subclassing additional hooks from this class. In this case, everything is (fairly boringly) some derivative of teapot.

This Hook class is not just used as your lowest-level interface with your service. It also describes how your connection variables are structured and should be presented to users. We implement a few special methods to make that happen:

```
@staticmethod
def get_connection_form_widgets():
    from flask_appbuilder.fieldwidgets import BS3TextFieldWidget
    from flask_babel import lazy_gettext
    from wtforms import StringField
    return {
        "pot_designator": StringField(lazy_gettext(
            "Pot Designator"), widget=BS3TextFieldWidget()),
        "additions": StringField(lazy_gettext(
            "Additions"), widget=BS3TextFieldWidget()),
    }
```

This method adds UI elements to the connections page of the Airflow webserver. Specifically, it will create two fields (Pot Designator and Additions) for our teapot connection type. It will store these fields as keys in the **extra** dictionary associated with a connection object:

```
@staticmethod
def get_ui_field_behaviour():
```

```
    return {
        "hidden_fields": ["password", "login", "schema", "extra"],
        "placeholders": {
            "pot_designator": "1",
            "additions": "sugar",
            "host": "tea-pot",
            "port": "8083",
        },
        "relabeling" : {}
    }
```

This modifies existing UI elements. Specifically, it will hide everything listed in `hidden_fields` and put placeholders in the listed elements as examples as someone is filling them in within the UI. The relabeling field is empty here and required, but you might use it to change the label on any preexisting field (for example, showing the schema field as protocol) in the UI while maintaining the schema attribute in the connection class. The authors generally don't suggest using this field as it decreases the overall understandability of your code:

```
@cached_property
def get_conn(self):
    conn = self.get_connection(self.tea_pot_conn_id)
    self.url = f"{conn.host}:{conn.port}"
    self.pot_designator = conn.extra_dejson.get(
        "pot_designator",None)
    self.additions = conn.extra_dejson.get("additions",None)
    return
```

This defines a method that fetches your connection from a database by ID and loads it into class attributes for later use. In this case, it simply builds your completed URL string based on host and port and sets your `pot_designator` and `additions` class attributes if they exist:

```
def test_connection(self):
    """Test a connection"""
    if self.is_ready():
        return (True, "Alive")
    return (False, "Not Alive")
```

This last method defines the `test_connection` method for your class. This will be executed every time the **Test** button from the Airflow Connections UI is triggered. In this method, we're relying on an unseen `is_ready` that calls the service's `ready` endpoint and checks for a good status.

We must also add one more key to our provider module's `get_provider_info` function return dictionary:

```
"connection-types" : [
             {
                 "connection-type": "teapot",
                 "hook-class-name": "airflow_provider_tea_pot.hooks.
 TeaPotHook"
             }
         ]
```

This tells Airflow that we are registering a connection type (teapot) that is described in the `hooks` module of our package with the `TeaPotHook` class. Now, when our provider is loaded through its entry point, the Airflow UI will also know about it in order to make adjustments to the connections UI page.

From here, you can write additional methods for interacting with your service:

```
    def __check_and_return(response):
        if response.status_code == 200:
            return response.text
        raise AirflowException(
            f"{response.status_code} : {response.reason}")

    def make_tea(self):
        self.get_conn
        response = requests.get(f"http://{self.url}/ready")
        return self.__check_and_return(response)

    def brew_coffee(self):
        self.get_conn
        response = requests.get(f"http://{self.url}/ready")
        return self.__check_and_return(response)

    def get_water_level(self):
        self.get_conn
        response = requests.get(f"http://{self.url}/ready")
        return self.__check_and_return(response)
```

In this example, we've gone with a very simple strategy of a method per exposed endpoint that includes a minimal check for a 200 HTTP response and raises an error otherwise. Your real-world use cases are likely to be much more involved and should include appropriate defensive practices and rigor to achieve your goals.

Authoring our operators

Now that we've completed our basic hooks, we can start working on our operators. For now, we're going to demonstrate a single operator for making tea:

```
class MakeTeaOperator(BaseOperator):

    template_fields = ()

    def __init__(self, tea_pot_conn_id, additions = None,
                 pot_designator = None, **kwargs) -> None:
        super().__init__(**kwargs)
        self.tea_pot_conn_id = tea_pot_conn_id
        self.pot_designator = pot_designator
        self.additions = additions


    def execute(self,context) -> typing.Any:
        self.hook = TeaPotHook(tea_pot_conn_id=self.tea_pot_conn_id)

        if self.pot_designator:
            self.hook.pot_designator = self.pot_designator
        if self.additions :
            self.hook.additions = self.additions
        return self.hook.make_tea()
```

Every operator requires two methods: an initialization method and an execute method. The initialization method should always be written to simply store configuration information for the instantiated operator and the execute method should be where you utilize your hook to get work done.

Authoring our sensor

Sensors are a special kind of operator that continually checks an external resource until they receive a signal that will cause them to succeed. Common use cases include event-based patterns where you don't have a scheduled time for something to occur, or if you need to ensure that a system is in a specific state before starting work.

Deferable Operators are a special kind of operator that put themselves into a special suspended state when needed. This state ensures that the Operators that are not actively doing work are able to give their worker slots back to the scheduler. When an Operator defers itself, it creates a trigger instance to run within a triggerer process. The trigger instance runs until the trigger event is met and the deferring operator is re-scheduled for completion with the registered callback.

Let's start by defining a trigger in the trigger's module. Your Trigger must inherent from the `BaseTrigger`, have an `__init__` method, have an asynchronous method called `run` that yields a `TriggerEvent`, and a `serialize` method that manages the trigger's ability to create itself:

```python
class WaterLevelTrigger(BaseTrigger):

    def __init__(self, tea_pot_conn_id, minimum_level) -> None:
        self.tea_pot_conn_id = tea_pot_conn_id
        self.minimum_level = minimum_level
        pass



    def serialize(self) -> typing.Tuple[
        str,typing.Dict[str,typing.Any]]:

        return "airflow_provider_tea_pot.triggers.WaterLevelTrigger", {
                    "minimum_level" : self.minimum_level,
                    "tea_pot_conn_id" : self.tea_pot_conn_id
                    }

    async def run(self):

        hook = TeaPotHook(tea_pot_conn_id=self.tea_pot_conn_id)
        async_get_water_level = sync_to_async(hook.get_water_level)

        while True:
            rv = await async_get_water_level()
            if json.loads(rv).get('level') > self.minimum_level :
                yield TriggerEvent(rv)
```

The code here describes a trigger that uses the TeaPotHook's `get_water_level` method to continually check that the water level of the teapot is above the minimum level. If it is, it will send a `TriggerEvent`. By default, the hook's `get_water_level` method is not asynchronous, so we use a convenience function from the `asgiref` package (`sync_to_async`) to make it an asynchronous call.

> **Trigger design considerations**
>
> Any blocking code within the run method must be asynchronous and use awaits whenever a blocking call is made.
>
> Trigger instances should be assumed to execute multiple times, so keeping your trigger idempotent is a best practice. You should at least be very mindful of the side effects associated with your run method.

Now that we have authored our trigger instance, we can write our deferrable operator. In this implementation, we're putting the operator in the sensors module because of how it interacts with the outside world:

```
class WaterLevelSensor(BaseOperator):

    def __init__(self, tea_pot_conn_id, minimum_level, **kwargs):
        super().__init__(**kwargs)
        self.tea_pot_conn_id = tea_pot_conn_id
        self.minimum_level = minimum_level


    def execute(self, context):
        self.defer(
            trigger=WaterLevelTrigger(
                tea_pot_conn_id=self.tea_pot_conn_id,
                minimum_level=self.minimum_level),
            method_name="execute_complete"
        )


    def execute_complete(self, context, event=None):
        return event
```

Our sensor is still just an operator, so it has an initialization method and an execute method. The execute method in this operator is simply a call to the `defer` method of the operator, and we tell it to defer using our newly defined trigger. When the trigger returns the `execute_complete` method is called. This section of code is where you would usually handle any additional business logic or checks that are needed for your operator.

Testing

As you are authoring your code, you should be writing test cases to help you validate your code's functionality and correctness. The corresponding code for this project has many test cases to help give ideas on how and what to test. One area we do want to spend a bit of time delving into is setting up a testing fixture so you can write more complete tests without needing to mock as many services.

In this example, we use `py.test` as our testing framework. There is a file in the `tests` folder, `conftest.py`, that `py.test` uses to provide fixtures for testing:

```
import os
os.environ["AIRFLOW__DATABASE__LOAD_DEFAULT_CONNECTIONS"] = "False"
os.environ['AIRFLOW__CORE__UNIT_TEST_MODE'] = 'True'
```

```
os.environ['AIRFLOW__CORE__LOAD_EXAMPLES'] = 'False'
os.environ['AIRFLOW_HOME'] = os.path.join(os.path.dirname(__
file__),'airflow')



@pytest.mark.filterwarnings("ignore:DeprecationWarning")
@pytest.fixture(autouse=True,scope='session')
def initdb():
    """Create a database for every testing session and add connections
to it."""

    from airflow.models import Connection
    from airflow.utils import db

    db.initdb(load_connections=False)

    db.merge_conn(
        Connection(...)# Add our connection information
    )

    yield
    #clean up behind ourselves
    shutil.rmtree(os.environ["AIRFLOW_HOME"])
```

This code does a few important things for you. When you run `py.test`, it will use this code to set some specific environment variables to configure Airflow for a test run, set the `AIRFLOW_HOME` directory to the testing directory, initialize the metadata backend (as a SQlite database file), and load a single connection to the database so it is available for your tests.

The `pytest.fixture` decorator is configured to run exactly one time for the entire testing session, yielding after setting up the database so that individual tests can execute. After the tests are complete, the code after the `yield` statement is run, cleaning up the environment.

Through the examples provided, you'll see fairly extensive use of mocked calls to external services; this is a requirement unless you wish to create additional fixtures to make actual services available for testing.

Functional examples

As a final step of authoring our provider, it's good practice to provide some functional examples of our DAG so we can demonstrate its function. To achieve this, we include a Dockerfile, `docker-compose.yaml`, and an `example_dags` directory in the project:

```
from datetime import datetime,timedelta
```

```
from airflow import DAG
from airflow_provider_tea_pot.operators import(
    MakeTeaOperator,BrewCoffeeOperator)
from airflow_provider_tea_pot.sensors import WaterLevelSensor

with DAG(...) as dag:


    t1 = WaterLevelSensor(
        task_id = "check_water_level",
        tea_pot_conn_id="tea_pot_example",
        minimum_level= 0.2
    )


    t2 = MakeTeaOperator(
        task_id = "make_tea",
        tea_pot_conn_id="tea_pot_example",
    )


    t3 = BrewCoffeeOperator(
        task_id = "brew_coffee",
        tea_pot_conn_id="tea_pot_example",
    )


t1 >> [t2,t3]
```

Our example DAG is simple. It runs once a day, we check the water level of our teapot, and if or when the level is high enough, it executes two operators to make tea and coffee in parallel:

```
ARG IMAGE_NAME="apache/airflow:2.5.0"
FROM ${IMAGE_NAME}

USER airflow
COPY --chown=airflow:airflow . ${AIRFLOW_HOME}/airflow-provider-tea-pot
COPY --chown=airflow:root example_dags/ /opt/airflow/dags
RUN pip install --no-cache-dir --user ${AIRFLOW_HOME}/airflow-provider-tea-pot/.
```

Our Dockerfile is pretty simple; it's the base Airflow image. We just copy our source code into the container, copy our example_dags folder into the default dags folder, and install the provider into the image.

For this DAG to work, we need a teapot to talk to. Obviously, no such teapot exists in real life. However, in the provided Docker Compose file, you'll see a `tea-pot` service. This is a simple HTTP server that provides endpoints for our fictitious teapot. When we run `docker compose up`, this service will come up along with a Postgres server and the core Airflow components for consumption. Once you have checked that all services are up and running by navigating to `localhost:8080`, you can log in to your Airflow instance and see (and even operate) your DAG.

Summary

In this chapter, you learned how to design a Python package that is structurally capable of integrating well with Airflow. You learned the basics of how to develop your testing program and even provide a functioning example for your users to consume. All you need to do now is distribute and support it!

In our next chapter, we will be creating a workflow for a machine learning use case. We will be starting with how to take a notebook from a data scientist, go through the design process, and end with a DAG that orchestrates the processing and loading of data along with compiling and distributing our model artifact.

8

Orchestrating a Machine Learning Workflow

In this chapter, we're going to go through an exercise in orchestrating processes to support a machine learning-based solution. We'll walk through the design and implementation of assets for a recommendation system. We'll cover some basic MLOps practices and show how to ensure our system is implemented in an operationally useful way.

In this chapter, we're going to cover the following topics:

- Basics of a machine learning-based project
- Designing your DAG
- Implementing your DAG
- Post-implementation activities

Technical requirements

In order to get the most out of this chapter, you should have completed earlier sections of this book and have a firm grasp of the basics of Airflow. All code is available within the GitHub repo for this book at the following URL: `https://github.com/PacktPublishing/Apache-Airflow-Best-Practices/tree/main/chapter-08`. We assume that you will have some basic knowledge of machine learning systems and ontologies, but an in-depth understanding of any specific technique or algorithm shouldn't be required.

Basics of a machine learning-based project

Machine learning is formally defined as the use and development of computer systems that learn and adapt, by using algorithms and statistical models to analyze and draw inferences from patterns in data. Systems that utilize these practices have some common functional components, including intaking and preparing data for use, applying a machine learning algorithm to the data to create a model, distributing the model artifact for use, and using the model to make inferences on new data.

> **Note**
>
> In addition to these components, there is usually a discussion of specialized telemetry collection relating to the performance of learned models and data. Those practices, while important, are well outside the scope of this book.

Airflow is a great tool for orchestrating portions of these common components – the intaking and processing of data, loading pre-computed features into backend services, the learning of models, and the distribution of model artifacts for consumption. Basically, it's ideal for everything in the "training" phase of the machine learning life cycle.

Airflow can also be useful in the "prediction" phase of the machine learning life cycle if your use case is for an offline batch system (e.g., getting predictions for a large list of observations on a well-known cadence). If your use case is for online predictions (e.g. serving predictions for a single observation at a time on demand), there are other, better, systems for this phase.

Our recommendation system – movies for you

We and a few friends are part of a large open community of movie buffs and reviewers, and we love to find ways to automate recommending movies to each other. One of our members happens to dabble in data science and has some experience building machine learning models to help with this.

Our data scientist friend plays around a bit with some sample datasets, and after some back and forth with the administrators and a few users to get feedback on the usefulness of a model, they believe that they have two models that, together, should provide the necessary data for recommendations to be useful.

After talking to the site administrators and working through details such as data access, privacy concerns, and general operational concerns, you agree on the following general constraints to the system:

- The website administrators will provide you with a series of files on a cadence that they will determine. They can't currently commit to a specific time frame because it's a volunteer website, and they aren't going to invest heavily in optimizing the data dump process. Also, they'll only run it when there are very few people online.

- The website wants to serve the models themselves for predictions, so you simply need to provide a system that updates the "vectors" of the KNN model on a regular cadence, along with the model training artifact (and metadata), to them for consumption.

> **Note**
>
> For this demonstration project, we'll use the "MovieLens" (`https://grouplens.org/datasets/movielens/`) dataset to provide us with real-world data.

Now that we know the basics of building an ML-based project, and have a project in mind, let's get started with designing the DAG for it.

Designing our DAG

As we review the code, we notice a few important things for each of the notebooks for these models; they're using identical code in a few sections to download and process data and for the "training" phase of the deep learning approach. After a little bit of time with a whiteboard and examining the code with your data scientist friend, you agree on the following design for your DAG.

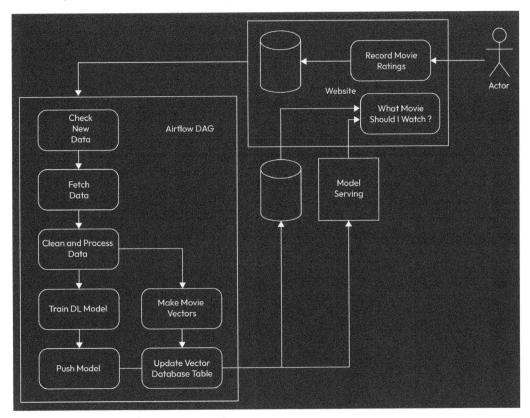

Figure 8.1: The design of our DAG

Here are some things you noticed during the design phase that will impact implementation details:

- Since you don't know exactly when data will become available for this process, it's very hard to utilize a purely scheduled approach; instead, we'll have to make sure our first step can determine whether the data has been changed before downloading and running all of these models.

- The notebooks your friend showed you had some repetitive code in them; you collapsed them into a single path in your DAG to simplify your processing and make it more consistent.

- Your friend was only using a small subset of the data during their initial development; the full dataset will require extensive compute resources to complete the work.

- When new data comes in, both movie and user vectors will be fundamentally different in structure and content. When we upload those vectors back to the website's database, we'll need to be able to "swap" data cleanly with minimal service interruption.

Now that we've got a high-level design and a series of technical requirements, we can move into implementation and testing.

Implementing the DAG

As we begin implementing our system with the aforementioned design, there are some basic capabilities within the system we're building that we'll need to determine how to implement.

Determining whether data has changed

When a new dataset is uploaded to a remote site, a secondary file that contains the md5 hash of the dataset is included. This md5 checksum will only change when the dataset we plan on downloading changes, so we'll begin our DAG by downloading and checking the contents of that file against a previously stored copy of that hash. If they don't match, we'll continue down the whole pipeline; otherwise, we'll gracefully exit and check again next week:

```
data_is_new = BranchPythonOperator(
    task_id = "data_is_new",
    python_callable=_data_is_new
)
def _data_is_new(ti, xcom_push=False, **kwargs):
    dataset_hash_location = __get_last_successful_hash()

    internal_md5 = Variable.get("INTERNAL_MD5", default_var=None)

    external_md5 = __get_external_hash(dataset_hash_location)
    if internal_md5 != external_md5 :
        ti.xcom_push(key='hash_id',
                    value=external_md5)
```

```
        return 'fetch_dataset'
    return 'do_nothing'
```

We achieved this with `BranchPythonOperator`; this operator uses the return value of a Python function to determine which task to run next. The function executed fetches the md5 sum from the downloaded file and checks it against the last successful run's md5 value, which we conveniently store in an Airflow variable. The function returns the next stages to be executed – either `do_nothing` or `fetch_dataset`.

Fetching data

From here, we download our dataset, unpack the relevant files needed, and lightly process them to ensure that we retain only the needed information for downstream use.

The `download_and_unpack` task fetches the ZIP archive that contains all of the files for this dataset, unzips them locally, and then uploads the required `ratings` and `movies` CSV files back to an object store. It may feel natural to immediately start processing data in the same task, but we generally separate the fetch and processing stages to ensure that if there are downstream failures, we only spend time downloading files one at a time:

```
    fetch_dataset = PythonOperator(
        task_id = "fetch_dataset",
        python_callable = _fetch_dataset
    )
 def __download_and_unpack_dataset(source):

    r = requests.get(source)
    r.raise_for_status()
    logging.info(f"successfully downloaded content from {source}")

    zip_dst = Path.home() / Path(source).name
    return_paths = []
    with open(zip_dst,"wb") as f:
        f.write(r.content)
```

After we've downloaded the ZIP file, we need to extract the relevant files from the archive and make sure to pass those file locations along for downstream processes to utilize:

```
    logging.info(f"External data downloaded to {zip_dst}")

    with zipfile.ZipFile(zip_dst,'r') as z:
        unzip_dst = Path.home()
        z.extractall(unzip_dst)
        Path(source).stem
```

```
        logging.info(f"Following files unzipped {z.namelist()}")

        # rebuild the absolute path to the files we need to process
        for f in z.namelist():
            if Path(f).name in ['ratings.csv', 'movies.csv']:
                abs_p = Path.home() / Path(f)
                return_paths.append(abs_p)

    logging.info(f"{return_paths} fetched and unpacked ")
    return return_paths
```

Note in the preceding code that we separate substantive code that does not interact with Airflow from code that primarily interacts with it. This provides a path for faster debugging and testing cycles with unit tests that don't require external services or extensive mocking. We also make sure to push an XCom value for the uploaded object's key so that downstream objects can determine where to access information from.

Pre-processing stage

We now process our CSV files into a tabular dataset that contains the vector representation of movies and users in our dataset. In a more involved pre-processing pipeline, this stage might involve multiple steps and many intermediate artifacts to ensure that mid-pipeline failures don't require extensive time to catch up:

```
    generate_data_frames = PythonOperator(
        task_id = "generate_data_frames",
        python_callable = _generate_data_frames
    )
def _generate_data_frames(ti, **kwargs):
    s3 = __get_s3_hook()
    bucket = __get_recsys_bucket()
    ratings_object = ti.xcom_pull(key="ratings.csv",
                                  task_ids="fetch_dataset")
    movies_object = ti.xcom_pull(key="movies.csv",
                                 task_ids="fetch_dataset")

    ratings_csv = s3.download_file(
        key = ratings_object,
        bucket_name = bucket
    )
    movies_csv = s3.download_file(
        key = movies_object,
        bucket_name = bucket
```

```
)
files_to_upload = _process_csv(ratings_csv=ratings_csv,
                               movies_csv=movies_csv)

for f, shape in files_to_upload:
    hash_id = ti.xcom_pull(key='hash_id',
                           task_ids="data_is_new")
    s3_dst = f"{hash_id}/{Path(f).name}"
    s3.load_file(
        filename=f,
        key=s3_dst,
        bucket_name= bucket,
        replace=True
        )

    ti.xcom_push(key=Path(f).name,
                 value=s3_dst)
    ti.xcom_push(key=f"{Path(f).name}.vector_length",
                 value = shape[1]-1)
```

Again, pay attention to the fact that we separate the "non-Airflow" sections of code away from the "Airflow" portions, aiding faster development iterations and unit testing without extensive mocking. We then upload to our object store and push XComs for both the object locations and some metadata for downstream use.

Once the data is prepared, we can parallelize downstream tasks that utilize it for independent tasks.

KNN feature creation

This portion of the pipeline deals with creating and storing the features for the collaborative filtering model. A keen observer will notice that the implementation here does not match the notebook too closely; in this implementation, instead of training a KNN model through `sklearn`, we insert the raw vectors into a specialized vector database that allows direct queries for similar matches:

```
enable_vector_extension = PostgresOperator(
    task_id="enable_vector_extension",
    postgres_conn_id=PG_VECTOR_BACKEND,
    sql="CREATE EXTENSION IF NOT EXISTS vector;",
)
load_movie_vectors = PythonOperator(
    task_id="load_movie_vectors",
    python_callable = _load_movie_vectors,
    op_kwargs={'pg_connection_id': PG_VECTOR_BACKEND},
)
```

```
        create_temp_table = PostgresOperator(
            task_id='create_temp_table',
            postgres_conn_id = PG_VECTOR_BACKEND,
            sql= 'DROP TABLE IF EXISTS "temp"; CREATE TABLE "temp" AS
TABLE "' + "{{ ti.xcom_pull(key='hash_id', task_ids='data_is_new') }}"
+ '";'
        )
def _load_movie_vectors(ti):
    s3 = __get_s3_hook()
    bucket = __get_recsys_bucket()
    hash_id = __get_current_run_hash(ti)
    vector_length = ti.xcom_pull(
        key='movie_watcher_df.parquet.vector_length',
        task_ids='generate_data_frames')
    movies_ratings_object = f"{hash_id}/movie_watcher_df.parquet"
    movie_ratings_file = s3.download_file(
        key = movies_ratings_object,
        bucket_name = bucket )
    # esablish postgres connection
    pg_hook = __get_pgvector_hook()
...

    def row_generator(df):
        for r in movie_ratings_df.rows():
            yield (r[0], f"{list(r[1:])}")
    movie_ratings_df = pl.read_parquet(movie_ratings_file)
    #bulk upload
    pg_hook.insert_rows(table=f'"{hash_id}"',
        rows=(r for r in row_generator(movie_ratings_df)),
        target_fields=['movieId','vector'])
    pass
```

This process consists of three steps – we ensure that the vector extension is enabled on the external database, we then load the data and vectors to a table that matches the hash of the current run, and then we create a second copy of that table and simply name it temp.

Notice that when we go to load the data to the vector table, we make defensive checks to drop the table if it already exists and then recreate it, using the metadata from the previous stage to inform how long it takes to set the vector length on the schema. We also utilize the insert_rows method of our postgres hook to ensure that our loads occur performantly in batches.

Our last stage is an accounting mechanism to set up the table for promotion. We make a full copy of the database and stage it under the table name temp so that when the deep learning model training is done, it's ready to quickly be promoted into production.

Deep learning model training

This portion of the pipeline deals with training ML algorithms. This section of the pipeline in a more serious use case is likely to require specialized compute infrastructure. To achieve this, we're going to use `KubernetesPodOperator` to offload compute from Airflow to a Kubernetes cluster. This is an incredibly powerful and flexible technique, as it allows you to use Kubernetes to schedule workloads to available compute flexibly and utilize node affinity to select for execution on specially configured hardware.

To complete this implementation, we have two major steps to complete – implementing our ML model as a standalone script and packaging that script into a Docker image for later use:

```
run_hash = os.environ.get('RUN_HASH',None)
...
#Get additional envirionment variables
def main():
    #download data into container

    # Prepare our dataset
    #Train the model
    #Save locally
    #Upload to the remote bucket
    #Push XCOMs
    #Kubernetes style.
    with open("/airflow/xcom/return.json", "w") as f:
        f.write(f'"{model_destination}"')
    #DockerOperator XCOM
    print(f"{model_destination}")
if __name__ == '__main__':
    main()
```

When implementing your script, you should design it knowing that configuration information comes in completely through environment variables or the command parameter of the Docker container. When choosing which avenue to use, we'd usually suggest defaulting to environment variables, as they're relatively easy to parse and can import secret values safely.

If you wish this script to push `xcom` at the end of its run back to Airflow for downstream use, you need to make sure you write a JSON-compatible text file to `/airflow/xcom/return.json`:

```
FROM python:3.11

WORKDIR /usr/src/app
```

```
RUN pip install --no-cache-dir numpy polars keras tensorflow scikit-
learn boto3 botocore && \
    mkdir -p /airflow/xcom && \
    echo "" > /airflow/xcom/return.json


COPY dags/recsys_dag/model_trainer.py .

CMD [ "python", "./model_trainer.py" ]
```

The Dockerfile is very straightforward – we use a basic python container as our base layer, install our necessary Python dependencies, create an empty file at /airflow/xcom/return.json, copy our script into the container, and set the entry point to execute the bundled Python script.

Once these two components are available, you can utilize the image within a Kubernetes cluster by using KubernetesPodOperator:

```
train_DL_model = KubernetesPodOperator(
    namespace="default",
    image="model_trainer",
    name="airflow-recsys-model-trainer",
    task_id="train_dl_model",
    do_xcom_push = True,
    in_cluster=False,
    env_vars = {
        'RUN_HASH' : "{{ ti.xcom_pull(key='hash_id',task_
ids='data_is_new')}}",
        'RECSYS_DATA_SET_KEY' : "{{ ti.xcom_pull(key='ratings.
csv',task_ids='fetch_dataset')}}"
    },
    cluster_context="docker-desktop",  # is ignored when in_
cluster is set to True
    config_file='/usr/local/airflow/include/.kube/config',
    on_failure_callback = _promotion_failure_rollback
)
```

> **Note**
>
> In order to make this code functional, you'll need to make sure that the kubeconfig file for the cluster you intend to use is available to your Airflow worker. You should obtain this file from the administrator of your Kubernetes cluster. This file should be treated with care and brought into your Airflow deployment, using whatever system you deem best to manage secrets.

Promoting assets to production

Note a NoOp step after both pipelines. This is an empty operator that purely serves as a checkpoint to ensure that the preceding pipelines complete successfully before starting any promotion processes. This is done to ensure that the promotion of both assets is done as close to simultaneously as possible, ensuring that the model and the feature store most likely present the same data to the application.

For this, we need to do two things:

- Promote the `temp` table to the `movie_vectors` table, which in this mock example is what the application uses to determine the vector of a particular movie.

- Copy the model training artifact to the `latest_model` key. We assume that this is sufficient for the application serving the model to manage reloading the asset for serving.

However, dealing with failures at this stage might be difficult, so we will implement error handlers to handle the rollback of both assets if either of them fail:

```
def _promotion_failure_rollback(context):
    ti = context['task_instance']
    run_hash = __get_last_successful_hash()

    s3 = __get_s3_hook()
    bucket = __get_recsys_bucket()

    pg_hook = __get_pgvector_hook()
    pg_hook.run(f"DROP TABLE IF EXISTS 'movie_vectors'; CREATE TABLE
'movie_vectors' AS TABLE '{run_hash}';")

    s3.copy_object(f"{run_hash}/model.keras", "latest_model.keras",
bucket, bucket)
```

This handler is registered as an on-failure callback for all tasks in the promotion process phase.

Finally, we set the MD5 hash variable to the current run's hash so that the next time the DAG runs, it can check that it is up to date against this dataset:

```
    update_internal_hash = PythonOperator(
        task_id = 'update_internal_hash',
        python_callable = _update_internal_hash,
        on_failure_callback = _promotion_failure_rollback
    )
def _update_internal_hash(ti,**kwargs):
    Variable.set("INTERNAL_MD5", __get_current_run_hash(ti))
    return
```

So now, we have a DAG that checks whether it needs to be run, fetches and pre-processes data, allows us to stage data for later promotion to production, allows us to execute remote code on specialized hardware to train a deep learning model, and safely executes the promotion of compiled artifacts (datasets and models) to production in a synchronized fashion. And all of it is done in an idempotent fashion so that even if a run is duplicated, it won't change the outcome!

On top of it all, because we were diligent about offloading intermediate states with a structured naming convention, we've got a way to go back and check historical and intermediate assets for investigation if we need to.

Summary

After completing this chapter, you should have the basics you might need to implement many different ML workflows in a way that is safe and auditable.

Keep in mind that this example is not meant to be fully inclusive of *everything* you might want or need to do to get a ML model into production, monitored, and ready for serving, but it should give you a good framework to start from and extend with additional requirements, based on your use case.

9

Using Airflow
as a Driving Service

Hopefully, by this point in your Airflow journey, the statement *Airflow is a capable tool for the definition of orchestration workflows*, is non-controversial. It should also be non-controversial to understand that, while powerful, it does require a more than cursory understanding of Python and Airflow internals to make it particularly valuable. At some point in your career, you will likely find a use case to abstract the authoring of workflows to a "less technical" group, allowing them to author and schedule workflows without having to have knowledge of Airflow or Python.

This chapter demonstrates a pattern in which Airflow is abstracted away from users. We do this to provide a simplified interface to Airflow that allows individuals to utilize Airflow to orchestrate and execute workloads without having to understand Airflow (or even know that it is responsible, as the backend service, for execution).

In this chapter, we are going to produce a hypothetical system for a QA engineering team. In this system, we're going to assume that our QA engineers will be using a hypothetical application to create a series of tests that can then be executed by Airflow.

In this chapter, we're going to cover the following topics:

- Defining the workflow we wish to provide as an abstracted service
- Choosing how to abstract our DAG definitions into something other than Python code
- How to utilize templating to allow for the creation of DAGs from templates
- How to ensure that are DAGs are appropriately scheduled for execution

Technical requirements

In order to complete this chapter, you should have completed earlier chapters of this book and have a firm grasp of the basics of Airflow. All code is available within the Git repo for this book at the following URL: `https://github.com/PacktPublishing/Apache-Airflow-Best-Practices/tree/main/chapter-09`.

We assume that you will have a complete understanding of basic Airflow components and functionalities.

QA testing service

We're the lead of a QA testing platform, and we have reason to believe that if we can make the configuration of end-to-end tests easier to define, we can increase the amount of testing we do and increase the overall quality of the software we deliver.

In order to achieve this, our solution will have a simple web UI that allows a user to describe a set of tests and, upon submission, Airflow will use those configurations to define a DAG run. The target state components and workflow of the total system can be seen in *Figure 9.1*.

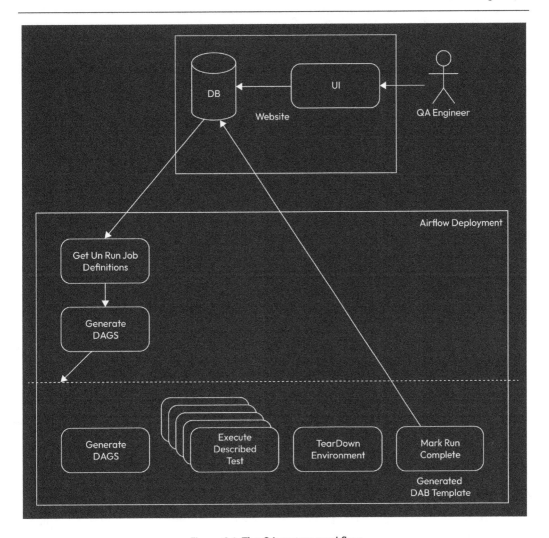

Figure 9.1: The QA system workflow

> **Note**
>
> For the purposes of this demonstration, we will not be providing the web frontend. Instead, we will only provide a mechanism that will create records and store them in what the application's backend could potentially look like.

Designing the system

Now that we have our use case and high-level system defined, we can go about defining some features of the system that are required for it to be successful:

- A definition for how we would like to describe workflows in a non-programmatic fashion that can be serialized and stored in a database
- A way to set up our testing environment before running described tests and tearing down that environment at the end of the run
- A way for taking described workflows and transpiling our DSL to a functional Airflow DAG
- A way to ensure that DAGs are unscheduled after they have successfully run

Choosing how to configure our workflows

For this system, we're going to assume that our tests can be abstracted as a repeatable function call with two arguments: name and value. (In the real world, this might correlate to an endpoint to be called and a return value or maximum runtime for a performance case.) Our UI will allow the engineers to define multiple test cases in a suite to be executed together. From a technical perspective, it will store the configurations as a JSON serialized list of dictionaries in a database:

```
[
  {"name": "test_1",  "value": 1 },
  {"name": "test_2", "value": 3  },
  {"name": "test_3", "value": 4  }
]
```

Defining our general DAG topology

As we alluded to earlier, each suite of tests needs to run in a special environment that needs to be created and configured before we execute our tests. Once that is complete, we can proceed to tear down the environment – we also want to make sure they tear down our environments upon exiting the DAG.

This behavior is probably familiar to anyone who writes involved unit or systems tests, and while completely achievable with some convoluted logic using nothing but TriggerRules in Airflow, as of the 2.7 release, Airflow provides some convenient methods to make this a bit easier.

As a starting point, we're also going to assume that our tests are isolated enough to be run independently of each other and don't require any sort of dependencies between each other. As such, the general topology of the DAG is simple: set up our environment; fan out tasks in parallel; on successful completion, notify the web service that the run was complete; and tear down the environment.

Creating our DAGs from our configurations

Now for the process with which we take our JSON description of the test suite and turn it into a DAG. For this, we're going to use the same templating engine that Airflow uses to render macros and templated variables, Jinja2:

```
#import statements
import datetime
from airflow import DAG
from airflow.operators.python import PythonOperator
#Python Functions
def _setup():
    pass
def _teardown():
    pass
def _test_case(s):
    pass
def _mark_success():
    pass
```

So far, the preceding code is in a format that should look very familiar to you. The DAG for running our test suite just imports code and defines some Python functions:

```
# DAG Definition
with DAG(
    {{ dag_id }},
    is_paused_upon_creation = False,
    start_date=datetime.datetime(2021, 1, 1),
    catchup=False,
    schedule="@once"
) as dag:
    setup_task = PythonOperator(_setup, ...)
    teardown_task = PythonOperator(_teardown,...)
    mark_successful = PythonOperator(_mark_success)
    tests = []
```

We start by creating our DAG context, creating a variable for the DAG ID. When we render an instance of this template, we'll need to ensure that the dag_id is a unique identifier for each run. In this case, we'll use the primary key from the database for a submitted test suite configuration. We also make sure that, upon creation, the DAG is unpaused. Upon creation, it has a start date that is sufficiently far back to guarantee an initial run and schedule of @once to ensure exactly one run:

```
    {% for task in tasks %}
    tests.append(
```

```
        PythonOperator(
            task_id = "{{ task.name }}",
            python_callable = _test_case,
            op_args = [{{ task.value }}]
        )
    )
    {% endfor %}
```

This segment of code is where we can actually define the more dynamic components of the DAG for templating. Variables to be substituted are marked with enclosing double curly brackets ({{ }}) and the name. Control structures are signified by using curly brackets and percent signs and are generally composed of statements that signify the beginning and end of the control statement. (For a full treatment on Jinja syntax, visit the Jinja website at https://jinja.palletsprojects.com/.)

In this example, we know we're going to be receiving a list of tasks and that we'll want to iterate over them with a for loop. The list we'll be receiving is a list of dictionaries with the two following keys: name and value. When the engine goes to render, it will print the text until the {% endfor %} statement for each iteration:

```
setup_environment.as_setup() >> tests >> teardown_environment.as_
teardown()
tests >> mark_successful
```

Finally, we'll make sure that we define the task dependencies for our entire graph. Notice that, in this example, we're setting the tests list as the upstream of both the teardown and mark successful tasks.

Now that we've defined our template, we need to understand how to render it:

```
with open("path/to/dag/folder","w") as f:
    env = Environment()
    template = env.from_string(dag_template)
    output = template.render(dict(
        dag_id = "unique_identifier",
        tasks = [{name:"test_one", value:1},
                 {name:"test_two", value:2}]
    ))
    f.write(output)
```

This bit of code opens a file to render our DAG file to, reads the defined template (all of the text we defined in the preceding), and then renders it using variables we pass in. In the preceding example, we'd print out a DAG with two defined tests to execute on. In our DAG writer loop that we'll discuss next, we'll be getting the values from dag_id and tasks from the backend database that stores test cases submitted by our QA engineers.

Scheduling (and unscheduling) our DAGs

You may be tempted to use a dynamic DAG pattern here – don't. While widely adopted, utilizing the scheduler to create DAGs and register them in a global namespace can be difficult to debug due to the ephemeral nature of the pattern. Also, our use case is going to be querying an external database. If that fails or is slow for any reason, it will negatively impact the scheduler's performance:

```
def _generate_dags():
    pg_hook = PostgresHook(postgres_conn_id=TEST_CASE_CONN_ID)
    results = pg_hook.get_records('SELECT case_id, case_descriptor
FROM "cases" WHERE "case_status" is NULL ')
    for r in results:
        with open(os.path.join("/opt","airflow","dags",
                            f"case_{r[0]}_dag.py"),"w") as f:
            env = Environment()
            template = env.from_string(dag_template)
            output = template.render(dict(
                dag_id = r[0],
                tasks = json.loads(r[1])
        ))
            f.write(output)
```

In order to schedule our test suites, we're going to have Airflow query the backend database for all test suites that have a Null status. Each record has a serialized JSON slug that can then be deserialized and used to render the **DAG to Airflow Is configured dag** folder. Once written, the scheduler loop will pick up and schedule the DAG for execution. Upon completion, the scheduled DAG will mark itself as "complete" in the backing services database.

Now, you might notice that there's a problem with the current setup: we're going to continually write DAGs to the folder with DAGs that are intended to run exactly one time. As the number of DAGs increases, the scheduler will get bogged down and you're likely to start seeing a degraded experience in executing DAGs:

```
def _drop_successful_dags():
    pg_hook = PostgresHook(postgres_conn_id=TEST_CASE_CONN_ID)
    results = pg_hook.get_records('SELECT case_id FROM "cases" WHERE
"case_status" = \'SUCCESS\'; ')
    for r in results:
        try:
            os.unlink(os.path.join("/opt","airflow","dags",
                            f"case_{r[0]}_dag.py"))
        except FileNotFoundError:
            pass
```

In order to handle this problem, we create a second task that queries the database for test suites marked "successful" and attempts to unlink (delete) DAGs that previously ran to completion. This ensures that the scheduler only processes and schedules DAGs that are still intended to be executed:

```
with DAG(
    "e2e_test_generator",
    start_date=datetime(2021, 1, 1),
    schedule_interval="*/2 * * * *",
    catchup=False,
) as dag:
    generate_dags = PythonOperator(
        task_id = "generate_dags",
        python_callable = _generate_dags
    )
    drop_successful_dags = PythonOperator(
        task_id = "drop_successful_dags",
        python_callable = _drop_successful_dags
    )
generate_dags
drop_successful_dags
```

Now that we have our two basic functions for both querying our backend database and writing templated DAGs and cleaning up DAGs that have completed execution, we can simply write a two-task DAG and schedule it to run on a cadence that we find appropriate.

Summary

With the completion of this chapter, you've got a system that can be used to allow "non-technical" users to author and execute workflows without having any understanding of Airflow (or even Python). This is a powerful pattern for abstraction that can allow you to give the power and stability of Airflow to users without forcing them to gain a deep understanding of the system.

Keep in mind that this implementation is meant to be illustrative of the pattern but is in no way complete for production. A good exercise would be to take the example code associated with this chapter and experiment with methods for authoring workflows with different levels of complexity, more functionality, alerts for the submitting QA engineer, and other Airflow operators entirely.

Be wary as you experiment with or adopt this pattern – taken to the extreme, you could find yourself attempting to define all of the capabilities in Airflow in a JSON object, which will only provide an abject lesson in futility.

In the next chapter, we'll be discussing the finer points of developing and operating your Airflow workflows in the real world.

Part 4:
Scale with Your
Deployed Instance

This part has the following chapters:

10

Airflow Ops:
Development and Deployment

In this chapter, we explore the application of modern **ops** practices within Apache Airflow, focusing on streamlining development and deployment workflows for your consuming teams.

We'll begin with a brief exploration and discussion of Airflow deployment methodologies, including Kubernetes, distributed virtual machines, and service providers.

With those foundations established, we'll dive into common patterns of the common DAG deployment methodologies: bundling, push, and pull. We'll then address repository structures, secrets management, and localized development.

No matter what path you decide on, it will generally require that you figure out methods to do a few common tasks:

- Customizing your Airflow deployment with additional operators, Python packages, and system packages
- Versioning your DAGs
- Delivering your DAGs to your Airflow deployment
- Managing Airflow configurations

Finally, we'll finish by discussing where, when, and what code to test within your Airflow deployment. This will equip you with the knowledge to optimize workflow management and ensure reliable and resilient orchestration deployments.

Technical requirements

By now, we expect you to have a good understanding of Airflow and its components, along with having used some sort of technology for CI/CD (Jenkins, GitLab Runners, GitHub Actions, Travis CI, and so on). We will not be discussing specific implementations within a CI/CD technology because there is no universally adopted system. Instead, we will simply describe patterns you may wish to consider.

We also assume at this point that you're using Docker (or some similar OCI-compliant) container system for describing how to build compute resources and have at least a conversational understanding of building and distributing container-based applications.

DAG deployments

We'll start with a discussion of how to get your Airflow DAGs to your Airflow deployment. We won't discuss specific implementation details because they can vary depending on your specific use case, security context, and operational needs.

Bundling

The DAG bundling pattern may be the most obvious pattern and is a very common starting point on your journey to using Airflow in production. It's also antithetical to the *code as configuration* paradigm that is core to Airflow's development and consumption.

The general pattern is to physically bundle your DAGs within the same file system as your Airflow system, even in a system with distributed compute. This is achieved by having your container build process install Airflow and copy your DAGs into the container during the image build process. This image can then be tagged and distributed to ensure that all components are operating using the exact same Airflow, plugin, and DAG versions.

This pattern is particularly powerful as it provides strong guarantees that all components are operating with identical copies of Airflow, providers, plugins, and DAGs. The downside of this pattern is that it results in relatively long build times, potentially large images, and relatively high amounts of downtime to stop, deploy, and restart all services in the Airflow deployment.

De-coupled DAG delivery

In this method, the deployment of the Airflow system and your Airflow DAGs are separated from each other so that they can proceed independently of each other.

The general pattern here is achieved by building your core Airflow deployment along with any additional Python or system packages and then deploying as a generally long-lived system. The DAGs are then delivered to the system by copying them to a file system that the Airflow system has mounted as a volume.

This pattern is powerful as it allows you to treat your Airflow deployment as a relatively long-lived and stable software deployment (assuming you're not changing the core system too often). The downsides of this pattern are that it requires you to maintain separate delivery processes for both your DAGs and Airflow system, adds additional overhead in managing requirements from your DAGs with your Airflow system, and comes with additional infrastructure management and selection for shared file storage.

Push versus pull

Within the decoupled DAG Delivery pattern, two basic patterns of delivery exist: push and pull.

In the push mechanism, during the CI/CD process, files are pushed from the source repository to the file system that the Airflow Deployment reads from. This pattern is conceptually straightforward and usually trivial to set up. However, in system recovery scenarios where Airflow's file system is being recovered, you will need to be aware to re-trigger the copy in order to complete recovery.

In the pull pattern, a process within the Airflow deployment is responsible for pulling data down from the external repository to the shared filesystem for consumption by the Airflow system. This pattern is generally preferable. While it does require additional system orchestration, it is easier to configure for automation once deployed. A common implementation of this pattern is to utilize the `git-sync` application to synchronize a Git repository holding your DAGs into the deployed system.

We generally suggest that the bundling approach be used early on in your adoption of Airflow while you may be making many changes to both the system and DAGs. At the same time, as you mature and the system stabilizes, you can adopt a de-coupled approach to ensure that your DAGs are the only quick moving portions of the deployment.

Once you've decided how you want to get your DAGs into your Airflow deployment, you can start designing your repository architectures depending on your organizational needs.

Repository structures

Repository structure is a seemingly trivial but deeply impactful decision in any project. Take your time to think about how your team currently works, what types of operational patterns you wish to support, and how your teams will interact with the structure before deciding. Remember – you can always change your mind!

Mono-repo

A monolithic repo is a repository pattern where all of the code bases are placed in a single repository and tracked (and released) under a single version history. For example, you might have separate folders that contain your Airflow build system, separate folders for in-house operators/plugins, and a folder for each team that is publishing and distributing DAGs.

The mono-repo pattern is very powerful when you're adopting a "working at the head" strategy. Even with multiple teams, it is very easy for all teams to have easy and direct access to all of the code required to make sure their DAGs run – just make sure they're working at the *head* of the history.

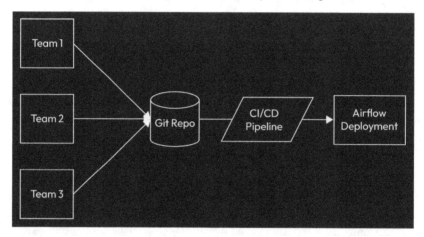

Figure 10.1: The mono-repo pattern

Major downsides of this pattern include increased repo size resulting in large download times, increased operational overhead in coordinating releases/promotions, and potentially more complex CI/CD pipelines when deploying as a microservice architecture.

Multi-repo

A multi-repo pattern is a repository pattern where you break up the entirety of your Airflow deployment into multiple repositories. Taken to its extreme, you might have a repo for core airflow, repos for each provider or plugin you use, and repos for each team deploying DAGs.

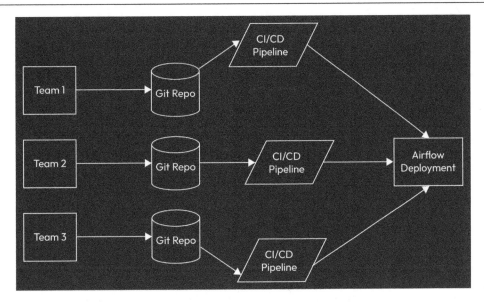

Figure 10.2: Multi-repo pattern

This pattern is great for allowing teams to have relatively simple and isolated workflows for the development and deployment of portions of the system they are responsible for. It does come at a cost, namely the synchronization of changes to the complete system across all teams for consumption. This is generally overcome with increased integration testing prior to release.

Mono-repo versus multi-repo is, of course, a false dichotomy. They are extremes on a spectrum to describe possibilities, and you will almost certainly wind up with a few small mono-repos. The authors would suggest the following:

- One mono repo responsible, distributed as a single container image, for describing your Airflow deployment and any in-house providers you are writing and a second set of repos for DAGs.

- It can also be a good idea to have multiple additional repos that allow teams to describe and deploy their DAGs. These teams can pull the upstream Airflow image for localized development and testing prior to release.

Once you and your team have had some discussions and made decisions about your intended repo structure, you can start making decisions about how to manage your Connection and Variable objects within your deployments. Make sure you include your security teams in these discussions as these objects generally contain sensitive information needed to connect to systems.

Connection and Variable management

Connection and Variable objects in Airflow are at the core of every operator and DAG. They're used to tell Airflow which systems it should be interfacing with and how a DAG should behave given the value in a Variable. They're used like configurations in any other piece of software. They may change between environments and are likely to contain secret information.

In many examples, you might see a step where the author shows you how to set up Connection and Variable objects from the WebUI. While this is fine for proof of concept examples, this is inherently a click-ops operation. It is subject to all of the operational issues associated with that class of actions, mainly the lack of a consistent method for managing configuration objects across multiple environments.

Luckily, Airflow has additional methods for provisioning these objects that integrate well with modern DevOps practices.

Environment variables

Both Connection and Variable objects can be defined in Airflow as environment variables:

```
export AIRFLOW_VAR_NAME=1
export AIRFLOW_VAR_NAME_JSON='{"hello":"world"}'
export AIRFLOW_CONN_MY_DATABASE='conn-type://login:password@host:port/
schema?param1=val1&param2=val2'
export AIRFLOW_CONN_MY_DATABASE='{
    "conn_type": "my-conn-type",
    "login": "my-login",
    "password": "my-password",
    "host": "my-host",
    "port": 1234,
    "schema": "my-schema",
    "extra": {
        "param1": "val1",
        "param2": "val2"
}}'
```

For both objects, you'll notice that the pattern is to prefix the ID for your object with either `AIRFLOW_CONN` or `AIRFLOW_VAR`. When starting up, Airflow will register the preceding objects and make them available in DAGs.

By its nature, this mechanism is not wholly secure if you're storing these variable definitions as raw text strings somewhere, so you'll usually pair this with an external secret management system that either stores the variables as cryptographically secure objects, decrypting them during deployment, or injects them from an external service at startup.

Secrets backends

The Secrets Backend is an Airflow adapter that allows you to pull secrets from a secrets backend instead of assuming you can utilize already existing backends from common providers such as AWS Secrets Manager, Google Cloud Secret Manager, Hashicorp Vault, Microsoft Azure Key Value, and Yandex Lockbox.

To configure this, simply set the `secrets` section of your Airflow configuration file with the `ClassName` of the provider you intend to use (make sure to install the provider) in `backend`, along with any of the `kwargs` needed to initialize the backend in `backend_kwargs`.

From here, any time you attempt to access a Variable or Connection, Airflow will check for the value in the `secrets` backend.

We strongly recommend that you talk with your security and DevOps teams about which options are available to you before making a decision on how you wish to provision these objects. This decision is one where you want to make sure you have the full support of your team to keep your data safe and secure.

Once you (with your security team's input) have decided the best way to manage your connection and variables for your deployments, you will need to decide on the best way to deploy your Airflow.

Airflow deployment methods

Your deployment methodology will largely depend on your organization's infrastructure capabilities and background. When selecting a methodology, select one you or someone in your organization can reasonably support given your budgets.

Kubernetes

Kubernetes makes the process of building and running complex applications (such as Airflow) much simpler. It does this by providing a simple declarative syntax for describing your application and abstractions for auxiliary requirements associated with operating multiple services that communicate with each other over the network.

With the availability of a community-supported and well-developed helm chart from the Apache Airflow project, the deployment of Airflow into a Kubernetes project is a relatively straightforward process.

If you already have familiarity with Kubernetes (either directly or as a service provided by another team), we strongly recommend this deployment method.

Virtual machines

If you are comfortable with bare metal, you might choose to deploy and distribute Airflow on virtual machines (or actually on bare metal computers). This path is not trivial and we would strongly recommend that you utilize a configuration management tool such as Terraform, Chef, Puppet, Ansible, Pulumi, or Cloud Formation to manage the provisioning and operation of infrastructure and software.

We won't go into any more detail than to say that if you think you're going to take this route, you should evaluate service providers, as unless there is a sunk cost in your compute infrastructure already, they're likely to be less expensive and more stable than rolling your own bare metal deployment.

Service providers

Service providers can be great for decreasing your cost of ownership and support burden in using Airflow. They do this by managing all of the underlying infrastructure provisioning and operations for you, promising you a life free of worry about both infrastructure and software operational burdens. Especially if you're unsure about operating your own infrastructure, you should carefully evaluate the providers that are available to you.

When evaluating a provider, you should still use this chapter. Providers have made very opinionated choices in how they wish to operate Airflow that will lock you into certain development techniques and patterns for deployment. These decisions may impact your team by making it difficult (or even impossible) to adopt new Airflow features, upgrade certain Python packages, or use certain community providers.

Make sure you do your research thoroughly on their suggestions for best practices and any tooling they provide to ensure that it makes sense for you!

Hopefully, at the end of these sections, you'll have made the decisions you need in order to implement your Airflow deployment. You may even have a running installation with a few small sample DAGs in it. These next phases of operation are valuable, hard, and often overlooked. They revolve around how to make sure that your developers can write and test DAGs efficiently.

Localized development

Modern CI/CD practices are all about failing fast. One of the best ways of doing that is to ensure that there is a local development environment available for doing basic tests against your code before running through more expensive workflows as part of your promotion process.

A localized Airflow environment usually takes one of two patterns: a Python virtual environment or Docker Compose. There are many fabulous and open source projects out there that provide environments for building and testing different development cycles within the Airflow ecosystem. We suggest that you check out `https://airflow.apache.org/ecosystem/#tools-integrating-with-airflow` for an up-to-date list of tools that may suit your needs.

Virtual environments

With a virtual environment-based development environment, you create a Python virtual environment that contains all of the Python packages and activates them prior to starting up your airflow components as individual processes within the environment. Usually, these will utilize a SQLite database as a backend and connect solely to external services that allow for it.

Especially in scenarios where you are just getting started or have very simple use cases, this can be a fast lightweight pattern. They do, however, tend to overlap with the local operating systems libraries and available services, so you may find yourself in situations where Airflow's needs are in direct conflict with other systems or processes.

Docker Compose

Docker Compose is a tool that uses a YAML specification to describe how a series of containers should be composed into a services stack for consumption. It is commonly used when you wish to localize a distributed system but still isolate it from the core operating system. Development systems utilizing this pattern will usually allow for greater flexibility in choosing what services to include as part of your development system and will more closely mirror your deployment environments at the cost of some increased initial complexity to get set up. They also generally require you to have greater local resources (CPU, RAM, and disk) available to operate more extensive service stacks.

Cloud development environments

These environments are very powerful and provide a best-in-class experience for developing Airflow. In this pattern, your developers can provision and customize service stacks in a cloud environment. They can then access them from their local computers for development and iteration. They're great because they allow for absolute elasticity in the infrastructure associated with your stack. However, these environments (even when provided as a service) are generally not as portable as the two previous examples and do take some additional investment to get set up, even when they are offered as a service.

Now that you've identified how you wish to go about virtualizing your services for localized development, we'll cover testing practices.

Testing

Testing may be one of the most controversial subjects in Airflow. In our opinion, this is likely due to people trying to take a dogmatic view of what is a pragmatic practice.

Put simply, testing should be about promoting trust that your deployment will operate in a consistent and predictable manner. While we will be giving some general advice (because we have scars that we'd prefer others not to have), we won't claim that this is a golden or even complete path to success. When things break or outages occur (and they will), part of your retrospective process should be to include any additional tests to ensure that the problem doesn't occur again in the future.

This pattern of test development means two things: your testing plan will inherently match your lived experience in operating Airflow and it will evolve over time as you mature.

Testing environments

Airflow is software that, due to its highly configurable nature, you will generally rebuild and release as an artifact before deploying it to an environment. As such, you will usually test it at a systems level for functionality before releasing it to your teams for consumption.

If you're using a bundled DAG pattern, you're in some luck. Any DAG-level tests you have can be run and tested alongside Airflow to ensure that you're not introducing any breaking changes.

If you've decoupled your DAGs from your Airflow system deployment, you will likely put your Airflow system through a full set of tests in an isolated pipeline before attempting to promote it into lower environments for DAG tests.

No matter what you do, it is advantageous to have at least one lower **sandboxed** environment that can be readily destroyed, recreated, and configured with current DAGs in operation. This will provide a way to validate the functionality of your system before it is released to support production workloads.

Testing DAGs

DAGs are configured as code, which really just means that they're software and should be treated as such. The authors generally suggest a testing scheme as follows:

- **Smoke tests**: An automated series of tests that essentially check that DAGs are well-defined and are valid Python code. You might also include any idiomatic checks for how DAGs are configured here.

- **Unit tests**: An automated series of tests that test any additional methods or functions that the authors have specifically written for their DAG. Usually, these tests come from PythonOperators or dynamic.

- **Functional tests**: These tests may or may not be automated depending on your needs and appetites. These tests should be documented regardless and consist of given a known system configuration (data in a database or state of an external system the DAG is interacting with), a DAG execution with a specific context or configuration, results in a specific outcome.

- **Performance tests**: These tests may or may not be automated depending on your needs. These tests run a series of workloads that have specific benchmarks (time to complete, memory consumption, CPU usage, and so on) that must be met for promotion.

The earlier tests in this scheme should be completed and passed before later tests are started. This is done because in the event a relatively inexpensive test fails, you don't want to commit to the more expensive tests.

These tests should be executed in environments that are appropriate for their context and for your company's operational guidelines. For example, smoke and unit tests might be completely self-contained within the CI/CD system execution environment, functional tests might be done in a QA environment where you have the freedom to edit data for individual test cases, and performance tests are done in a staging environment that is a near 1:1 match of production. You should talk with your infrastructure, security, and/or DevOps teams to design a testing strategy that meets your needs and the company's standards.

Testing providers

Provider testing is generally less onerous than DAG testing as the breadth of functional testing is usually less extensive and easier to automate.

- **Smoke tests**: Test that the Python code you've written is functional. These are usually as simple as *can I install the package?*.

- **Unit tests**: These basically test the smallest units of code you can that don't require external services to work correctly. You'll usually fake connections to external services with a testing framework that provides a mocking interface.

- **Functional/integration tests**: These involve testing against an actual running service validating that the code works as expected against the actual systems required. These tests will usually include managing the creation and destruction of databases or other services required to do tests, along with preloading any data if needed.

Testing Airflow

The short answer here is "don't," unless you've materially changed the Airflow codebase and are maintaining a separate fork. (You should only do this under the direst of circumstances.)

Airflow is extensively tested upstream by the maintainers of the project. If you're testing core Airflow in any way for correctness as part of your deployment, you should take a step back and reevaluate why you're taking these actions and whether you should reach out to the community for help.

Summary

Incorporating CI/CD practices into Apache Airflow workflows empowers teams to streamline development, improve reliability, and accelerate time-to-market. By carefully selecting deployment methodologies, establishing robust repository structures, and prioritizing testing, organizations can harness the full potential of Airflow for orchestrating complex data pipelines in modern data-driven environments.

A keen observer will notice that we're missing a major segment of any ops-minded practice – monitoring and observability of the system itself! Don't worry; we'll be covering all the things you do once you're in production in the next chapter.

11

Airflow Ops Best Practices: Observation and Monitoring

In this chapter, we will continue to explore the application of modern "ops" practices within Apache Airflow, focusing on the observation and monitoring of your systems and DAGs after they've been deployed.

We'll divide this observation into two segments – the core Airflow system and individual DAGs. Each segment will cover specific metrics and measurements you should be monitoring for alerting and potential intervention.

When we discuss monitoring in this section, we will consider two types of monitoring – active and suppressive.

In an active monitoring scenario, a process will actively check a service's health state, recording its state and potentially taking action directly on the return value.

In a suppressive monitoring scenario, the absence of a state (or state change) is usually meaningful. In these scenarios, the monitored application sends an active schedule to a process to inform it that it is OK, usually suppressing an action (such as an alert) from occurring.

This chapter covers the following topics:

- Monitoring core Airflow components
- Monitoring your DAGs

Technical requirements

By now, we expect you to have a good understanding of Airflow and its core components, along with functional knowledge in the deployment and operation of Airflow and Airflow DAGs.

We will not be covering specific observability aggregators or telemetry tools; instead, we will focus on the activities you should be keeping an eye on. We strongly recommend that you work closely with your ops teams to understand what tools exist in your stack and how to configure them for capture and alerting your deployments.

Monitoring core Airflow components

All of the components we will discuss here are critical to ensuring a functioning Airflow deployment. Generally, all of them should be monitored with a bare minimum check of *Is it on?* and if a component is not, an alert should surface to your team for investigation. The easiest way to check this is to query the REST API on the web server at `` `/health/` ``; this will return a JSON object that can be parsed to determine whether components are healthy and, if not, when they were last seen.

Scheduler

This component needs to be running and working effectively in order for tasks to be scheduled for execution.

When the scheduler service is started, it also starts a `` `/health` `` endpoint that can be checked by an external process with an active monitoring approach.

The returned signal does not always indicate that the scheduler is working properly, as its state is simply indicative that the service is up and running. There are many scenarios where the scheduler may be operating but unable to schedule jobs; as a result, many deployments will include a canary dag to their deployment that has a single task, acting to suppress an external alert from going off.

Import metrics that airflow exposes for you include the following:

- `scheduler.scheduler_loop_duration`: This should be monitored to ensure that your scheduler is able to loop and schedule tasks for execution. As this metric increases, you will see tasks beginning to schedule more slowly, to the point where you may begin missing SLAs because tasks fail to reach a schedulable state.

- `scheduler.tasks.starving`: This indicates how many tasks cannot be scheduled because there are no slots available. Pools are a mechanism that Airflow uses to balance large numbers of submitted task executions versus a finite amount of execution throughput. It is likely that this number will not be zero, but being high for extended periods of time may point to an issue in how DAGs are being written to schedule work.

- `scheduler.tasks.executable`: This indicates how many tasks are ready for execution (i.e., queued). This number will sometimes not be zero, and that is OK, but if the number increases and stays high for extended periods of time, it indicates that you may need additional computer resources to handle the load. Look at your executor to increase the number of workers it can run.

Metadata database

The metadata database is used to store and track all of the metadata for your Airflow deployments' previous DAG/task executions, along with information about your environment's roles and permissions. Losing data from this database can interrupt normal operations and cause unintended consequences, with DAG runs being repeated.

While critical, because it is architecturally ubiquitous, the database is also least likely to encounter issues, and if it does, they are absolutely catastrophic in nature.

We generally suggest you utilize a managed service for provisioning and operating your backing database, ensuring that a disaster recovery plan for your metadata database is in place at all times.

Some active areas to monitor on your database include the following:

- **Connection pool size/usage**: Monitor both the connection pool size and usage over time to ensure appropriate configuration, and identify potential bottlenecks or resource contention arising from Airflow components' concurrent connections.

- **Query performance**: Measure query latency to detect inefficient queries or performance issues, while monitoring query throughput to ensure effective workload handling by the database.

- **Storage metrics**: Monitor the disk space utilization of the metadata database to ensure that it has sufficient storage capacity. Set up alerts for low disk space conditions to prevent database outages due to storage constraints.

- **Backup status**: Monitor the status of database backups to ensure that they are performed regularly and successfully. Verify backup integrity and retention policies to mitigate the risk of data loss if there is a database failure.

Triggerer

The Triggerer instance manages all of the asynchronous operations of deferrable operators in a deferred state. As such, major operational concerns generally relate to ensuring that individual deferred operators don't cause major blocking calls to the event loop. If this occurs, your deferrable tasks will not be able to check their state changes as frequently, and this will impact scheduling performance.

Import metrics that airflow exposes for you include the following:

- `triggers.blocked_main_thread`: The number of triggers that have blocked the main thread. This is a counter and should monotonically increase over time; pay attention to large differences between recording (or quick acceleration) counts, as it's indicative of a larger problem.

- `triggers.running`: The number of triggers currently on a triggerer instance. This metric should be monitored to determine whether you need to increase the number of triggerer instances you are running. While the official documentation claims that up to tens of thousands of triggers can be on an instance, the common operational number is much lower. Tune at your discretion, but depending on the complexity of your triggers, you may need to add a new instance for every few hundred consistent triggers you run.

Executors/workers

Depending on the executor you use, you will need to monitor your executors and workers a bit differently.

The Kubernetes executor will utilize the Kubernetes API to schedule tasks for execution; as such, you should utilize the Kubernetes events and metrics servers to gather logs and metrics for your task instances. Common metrics to collect on an individual task are CPU and memory usage. This is crucial for tuning requests or mutating individual task resource requests to ensure that they execute safely.

The Celery worker has additional components and long-lived processes that you need to metricize. You should monitor an individual Celery worker's memory and CPU utilization to ensure that it is not over- or under-provisioned, tuning allocated resources accordingly. You also need to monitor the message broker (usually Redis or RabbitMQ) to ensure that it is appropriately sized. Finally, it is critical to measure the queue length of your message broker and ensure that too much "back pressure" isn't being created in the system. If you find that your tasks are sitting in a queued state for a long period of time and the queue length is consistently growing, it's a sign that you should start an additional Celery worker to execute on scheduled tasks. You should also investigate using the native Celery monitoring tool Flower (`https://flower.readthedocs.io/en/latest/`) for additional, more nuanced methods of monitoring.

Web server

The Airflow web server is the UI for not just your Airflow deployment but also the RESTful interface. Especially if you happen to be controlling Airflow scheduling behavior with API calls, you should keep an eye on the following metrics:

- **Response time**: Measure the time taken for the API to respond to requests. This metric indicates the overall performance of the API and can help identify potential bottlenecks.

- **Error rate**: Monitor the rate of errors returned by the API, such as 4xx and 5xx HTTP status codes. High error rates may indicate issues with the API implementation or underlying systems.

- **Request rate**: Track the rate of incoming requests to the API over time. Sudden spikes or drops in request rates can impact performance and indicate changes in usage patterns.

- **System resource utilization**: Monitor resource utilization metrics such as CPU, memory, disk I/O, and network bandwidth on the servers hosting the API. High resource utilization can indicate potential performance bottlenecks or capacity limits.

- **Throughput**: Measure the number of successful requests processed by the API per unit of time. Throughput metrics provide insights into the API's capacity to handle incoming traffic.

Now that you have some basic metrics to collect from your core architectural components and can monitor the overall health of an application, we need to monitor the actual DAGs themselves to ensure that they function as intended.

Monitoring your DAGs

There are multiple aspects to monitoring your DAGs, and while they're all valuable, they may not all be necessary. Take care to ensure that your monitoring and alerting stack match your organizational needs with regard to operational parameters for resiliency and, if there is a failure, recovery times. No matter how much or how little you choose to implement, knowing that your DAGs work and if and how they fail is the first step in fixing problems that will arise.

Logging

Airflow writes logs for tasks in a hierarchical structure that allows you to see each task's logs in the Airflow UI. The community also provides a number of providers to utilize other services for backing log storage and retrieval. A complete list of supported providers is available at `https://airflow. apache.org/docs/apache-airflow-providers/core-extensions/logging.html`.

Airflow uses the standard Python logging framework to write logs. If you're writing custom operators or executing Python functions with a PythonOperator, just make sure that you instantiate a Python logger instance, and then the associated methods will handle everything for you.

Alerting

Airflow provides mechanisms for alerting on operational aspects of your executing workloads that can be configured within your DAG:

- **Email notifications**: Email notifications can be sent if a task is put into a marked or retry state with the `email_on_failure` or `email_on_retry` state, respectively. These arguments can be provided to all tasks in the DAG with the `default_args` key work in the DAG, or individual tasks by setting the keyword argument individually.

- **Callbacks**: Callbacks are special actions that are executed if a specific state change occurs. Generally, these callbacks should be thoughtfully leveraged to send alerts that are critical operationally:

  - `on_success_callback`: This callback will be executed at both the task and DAG levels when entering a successful state. Unless it is critical that you know whether something succeeds, we generally suggest not using this for alerting.

  - `on_failure_callback`: This callback is invoked when a task enters a failed state. Generally, this callback should always be set and, in critical scenarios, alert on failures that require intervention and support.

  - `on_execute_callback`: This is invoked right before a task executes and only exists at the task level. Use sparingly for alerting, as it can quickly become a noisy alert when overused.

  - `on_retry_callback`: This is invoked when a task is placed in a retry state. This is another callback to be cautious about as an alert, as it can become noisy and cause false alarms.

- `sla_miss_callback`: This is invoked when a DAG misses its defined SLA. This callback is only executed at the end of a DAG's execution cycle so tends to be a very reactive notification that something has gone wrong.

SLA monitoring

As awesome of a tool as Airflow is, it is a well-known fact in the community that SLAs, while largely functional, have some unfortunate details with regard to implementation that can make them problematic at best, and they are generally regarded as a broken feature in Airflow. We suggest that if you require SLA monitoring on your workflows, you deploy a CRON job monitoring tool such as healthchecks (`https://github.com/healthchecks/healthchecks`) that allows you to create suppressive alerts for your services through its rest API to manage SLAs. By pairing this third-party service with either HTTP operators or simple requests from callbacks, you can ensure that your most critical workflows achieve dynamic and resilient SLA alerting.

Performance profiling

The Airflow UI is a great tool for profiling the performance of individual DAGs:

- **The Gannt chart view**: This is a great visualization for understanding the amount of time spent on individual tasks and the relative order of execution. If you're worried about bottlenecks in your workflow, start here.

- **Task duration**: This allows you to profile the run characteristics of tasks within your DAG over a historical period. This tool is great at helping you understand temporal patterns in execution time and finding outliers in execution. Especially if you find that a DAG slows down over time, this view can help you understand whether it is a systemic issue and which tasks might need additional development.

- **Landing times**: This shows the delta between task completion and the start of the DAG run. This is an un-intuitive but powerful metric, as increases in it, when paired with stable task durations in upstream tasks, can help identify whether a scheduler is under heavy load and may need tuning.

Additional metrics that have proven to be useful (but may need to be calculated) include the following:

- **Task startup time**: This is an especially useful metric when operating with a Kubernetes executor. To calculate this, you will need to calculate the difference between `start_date` and `execution_date` on each task instance. This metric will especially help you identify bottlenecks outside of Airflow that may impact task run times.

- **Task failure and retry counts**: Monitoring the frequency of task failures and retries can help identify information about the stability and robustness of your environment. Especially if these types of failure can be linked back to patterns in time or execution, it can help debug interactions with other services.

- **DAG parsing time**: Monitoring the amount of time a DAG takes to parse is very important to understand scheduler load and bottlenecks. If an individual DAG takes a long time to load (either due to heavy imports or long blocking calls being executed during parsing), it can have a material impact on the timeliness of scheduling tasks.

Summary

In this chapter, we covered some essential strategies to effectively monitor both the core Airflow system and individual DAGs post-deployment. We highlighted the importance of active and suppressive monitoring techniques and provided insights into the critical metrics to track for each component, including the scheduler, metadata database, triggerer, executors/workers, and web server. Additionally, we discussed logging, alerting mechanisms, SLA monitoring, and performance profiling techniques to ensure the reliability, scalability, and efficiency of Airflow workflows. By implementing these monitoring practices and leveraging the insights gained, operators can proactively manage and optimize their Airflow deployments for optimal performance and reliability.

In the next chapter, we'll begin discussing multi-tenancy in Airflow and how it can be achieved in different ways.

12
Multi-Tenancy in Airflow

As your organization increases its adoption of Airflow, it is likely that more people and teams are going to want to use it. This increased usage means that there will be an increased load on what is a fundamentally shared architecture, so figuring out how to safely support different groups and individuals with different needs effectively is a critical step in your maturation journey.

In this chapter, we'll discuss options and configurations to isolate workloads and share infrastructure in Airflow. Multi-tenancy can mean different things to different people, and we should start by plainly stating that, by many definitions, Airflow is not a multi-tenant application. Many of the patterns and practices presented in this chapter will help you provide for a shared environment but will not necessarily provide high levels of isolation and security.

In this chapter, we'll cover the following areas:

- When to (and not to) choose multi-tenancy strategies in Airflow
- Common operational additions needed to support multi-tenant usage of Airflow

By the end of this chapter, you should have a solid understanding of when and why to choose a multi-tenant strategy along with operational activities you should undertake to make sure your multi-tenant Airflow is as secure as possible.

Technical requirements

At this point, we assume you have a strong understanding of the components that make up Airflow, along with how DAGs are generally developed and delivered to their executing environment. The code for this chapter is available at `https://github.com/PacktPublishing/Apache-Airflow-Best-Practices/tree/main/chapter-12`.

When to choose multi-tenancy

Multi-tenancy can mean many things to many people and for the duration of this chapter we're going to use the working definition of *multiple individuals with different needs utilizing a shared Airflow instance*. This wish usually comes from a need to ensure that the organization is getting the most out of its Airflow deployment, ensuring that it is being used to its utmost potential.

The choice to run Airflow in a multi-tenant manner should not be made lightly, and we have included this chapter partially as a warning that it should be avoided. With the rise of Kubernetes, community-supported Helm charts, and paid Airflow services, it is generally preferable to operate an Airflow per organizational unit (team, business, or security group) than attempt to manage large numbers of conflicting needs within a single deployment.

If you find yourself in a situation where you cannot avoid sharing a deployment, we'll present some technical guidelines in the next section to help you achieve your goals. But, keep in mind that there are likely not going to be any sort of strong policy-based guarantees to promote strong isolation between user interactions and workloads; you will instead rely on convention and governance to maintain safety.

Component configuration

All of the components within Airflow will be shared by all users (and DAGs) of a deployment. There are ways to potentially use operational paradigms and conventions to isolate information and provide a semblance of security, but these will likely not be able to be strictly enforced as a policy.

The Celery Executor

The Celery Executor launches multiple threads that pick up work from the broker for execution. By design, each thread shares CPU, memory, and local disk space within the worker, so neighboring workloads have a possibility of colliding and clobbering each other. If you wish to direct workloads to specific workers (or isolate them), you can utilize the queueing mechanism to do so. To do so, you need to ensure that each worker starts with a queue name that it is listening to (`airflow celery worker -q queue_name`). You can then assign tasks to specific queues by assigning the queue name to the operator using the `queue` keyword argument.

Each worker can then be configured with its own localized security policies to control.

To enforce queue assignment, you could utilize tools such as linters or pre-commit hooks to enforce any sort of queue naming and selection policy prior to DAGs being promoted into the execution environment.

The Kubernetes executor

The Kubernetes executor is highly isolated. Each process runs within its own Pod. If workload isolation is of extreme importance to you, this executor will likely give you the simplest and most security with the least amount of configuration required. If you're looking for specialized security contexts, utilizing a Pod override and either node selector taints or images are your best options.

The scheduler and triggerer

With current Airflow architectures, these components cannot be made multi-tenant in any way. It will always have full access to the DAGs and be responsible for scheduling all DAGs within the DagBag. If, for any reason, you need isolation within these components, your only option is to operate additional Airflow deployments.

DAGs

DAGs are available within a single directory of Airflow, and because of Airflow's (and Python's) inner workings are inherently loaded into the same namespace at runtime. This means that, while are not explicitly aware of each other, information within a DAG is implicitly available from any neighboring DAG using core Python import utils and basic file system access during parsing. There is functionally no way to support any sort of isolation within the core DAG folder, but when utilizing "push" or "pull" methodologies, it is relatively easy to deliver a disparate and distributed set of DAGs to a single Airflow instance for scheduling.

Web UI

The web UI provides multi-tenancy through roles and permissions. Users can be provisioned with roles that allow permissions to access Airflow objects. Let's explore adding custom roles and permissions to accomplish multi-tenancy in the Airflow web UI.

Permissions

Permissions in Airflow consist of paired nouns and verbs where the nouns are standard Airflow objects (DAG, DagRun, Task, Connection, Variable) and the actions are things such as create, read, edit, and delete.

Permissions are then applied to roles and endpoints, and when a user with a specific role attempts to access an endpoint, they are allowed to utilize it if they meet all required permissions for the resource.

Roles

The Airflow UI (and REST API) support access control at a variety of levels with a series of default roles that allow increasing levels of interaction and execution permissions within the environment. Airflow provides the following roles "out of the box" for you to utilize:

- Public
- Viewer
- User
- Op
- Admin

You can additionally create custom roles and bind them to user accounts based on your organizational needs to provide fine-grained access down to specific DAGs. A full list of resources and actions can be found in the Airflow source code within the `security.permission` module (`https://github.com/apache/airflow/blob/main/airflow/security/permissions.py`).

Once you have these basics in place, the process to create a custom role and assign it permissions is pretty straightforward. In this example, we're going to assume that you have two users who each have a separate folder that contains their DAGs and we want to make sure that they can't access each other's DAGs from the web UI. We'll control our behavior from an "administrative" DAG that runs in a separate tenant to ensure that permissions are synchronized regularly:

```
def __add_users(user_name, password, email, first_name ='', last_
name='', role='Public'):
    with get_application_builder() as appbuilder:
        role = appbuilder.sm.find_role(role)
        if appbuilder.sm.find_user(user_name):
            print(f"{user_name} already exist in the db")
            return

        user = appbuilder.sm.add_user(
            user_name, first_name, last_name,
            email, role, password
        )

def _add_users():
    users = [
        ("user_1", "user_1", "user_1@test.com"),
        ("user_2", "user_2", "user_2@test.com"),
    ]
    for u in users:
        __add_users(*u)
```

In this snippet of code, we create a user using the core Airflow libraries. (You could also do this with the command-line interface via `airflow users create`, but we've chosen to show you a programmatic option instead.) We instantiate our users with the `Public` role, which provides no real access. While, in this example, we do this within a DAG, you will more likely want to find a more secure way of managing your user creation using one of the available authentication plugins available:

```
    for d in os.listdir("dags"):
        if d != "admin_dags":
            with get_application_builder() as appbuilder:
                appbuilder.sm.add_role(d)
                role = appbuilder.sm.find_role(d)
                perm: Permission | None = appbuilder.sm.create_
permission(READ, "Website")
                appbuilder.sm.add_permission_to_role(role, perm)
```

```
                print(f"Added {perm} to role {d}")
                for r in [
                    "Task Instances", "Website", "DAG Runs",
                    "Audit Logs", "ImportError", "XComs",
                    "DAG Code", "Plugins", "My Password",
                    "My Profile", "Jobs", "SLA Misses",
                    "DAG Dependencies", "Task Logs"] :
                    perm: Permission | None = appbuilder.sm.create_
permission(READ,r)
                    appbuilder.sm.add_permission_to_role(
                        role, perm)

                for r in ["Task Instances", "My Password",
                        "My Profile", "DAG Runs"]:
                    perm: Permission | None = appbuilder.sm.create_
permission(EDIT,r)
                    appbuilder.sm.add_permission_to_role(
                        role, perm)
                for r in ["DAG Runs", "Task Instances"]:
                    perm: Permission | None = appbuilder.sm.create_
permission(CREATE,r)
                    appbuilder.sm.add_permission_to_role(
                        role, perm)
                for r in ["DAG Runs", "Task Instances"]:
                    perm: Permission | None = appbuilder.sm.create_
permission(DELETE,r)
                    appbuilder.sm.add_permission_to_role(
                        role, perm)
                for r in [
                    "View Menus", "Browse", "Docs",
                    "Documentation", "SLA Misses", "Jobs",
                    "DAG Runs", "Audit Logs", "Task Instances",
                    "DAG Dependencies"]:
                    perm: Permission | None = appbuilder.sm.create_
permission(MENU_ACCESS,r)
                    appbuilder.sm.add_permission_to_role(
                        role, perm)
```

Next, we add the role needed to the user and then make sure to assign that role to the user. This block of code simply adds the minimum permissions a user might need to generally access. Notice how the DAG object is largely absent:

```
                dag_bag = DagBag(os.path.join("dags",d))
```

```
        for action in [READ,EDIT,CREATE,DELETE]:
            for d_id in dag_bag.dags.keys():
                perm: Permission | None = appbuilder.
sm.create_permission(action, f"DAG:{d_id}")
                appbuilder.sm.add_permission_to_role(
                    role, perm)
                print(f"Added {perm} to role {d}")
            user = appbuilder.sm.find_user(d)
            user.roles.append(role)
            appbuilder.sm.update_user(user)
```

Finally, we instantiate a DagBag object on the user's DAGs (conveniently, the folder name and username are the same) in order to get a list of the DAG IDs. We then apply a special permission in the format of DAG:<dag_id> to ensure that this role has full access to the DAGs in their folder. Finally, we make sure to assign that role to the same user.

Summary

Airflow was designed as an application that has a fairly open and shared architecture; however, with additional effort, you can isolate information between users to make it a multi-tenant environment. There are multiple strategies that rely on architectural and configuration options available to you. While no single option is likely to cover your needs, we hope that a combination of the presented options will help you achieve your security and operational goals. Remember, there is always an option to fully isolate with an additional Airflow deployment!

In the next chapter, we'll discuss valuable strategies for planning and executing Airflow migrations.

13

Migrating Airflow

In the data engineering profession, you're likely going to need to migrate something every few years due to reasons such as new technology selections, infrastructure changes, or disaster recovery. No matter what the reason for carrying out a migration that goes poorly, they generally are deeply impactful to business processes and operations. Taking time to plan and de-risk associated activities is critical for a successful migration.

In this chapter, we'll cover the following areas:

- Work management activities to plan and execute a pipeline migration
- Technical approaches to do your migration.
- Specific methods to migrate from one Airflow instance to another

By the end of this chapter, you will have an understanding of the activities required to plan for and execute a successful migration.

Technical requirements

At this point, you should have a thorough understanding of the inner workings of Airflow and the general operational patterns associated with the movement and organization of data.

General management activities for a migration

No matter what technologies you're migrating to or from, there are some common activities that are required in order to ensure that your migration is executed in a way that is systematic, repeatable, and as safe as possible.

Inventory

As obvious as it may seem, you need to start with a full inventory of the workflows that you are going to migrate. Start with a list of all of the workflows and identify the following information – who owns the workflow from a technical perspective, who owns it from a business/operational perspective, what data sources and destinations it interacts with, and how critical it is from a business perspective.

As you're compiling this information, make sure you can answer the following questions for each workflow:

- Where does the code/configuration for this workflow reside?
- How often does the workflow run?
- What edge cases/tests are compiled for this workflow?
- What does "right" look like when the workflow is running correctly?
- How has the workflow broken in the past?
- How big of a deal is it if this workflow is off for a period of time or running improperly?

Sequence

Once you've got a full inventory, start breaking out groups of workflows into tranches of work to be completed. A common pattern would be to organize them into groups by operational or technical units, and then prioritize them from lowest to highest based on the business criticality of the migration:sequence" workflows. Take your time here, as you're likely to find some hidden work and information as you go through this stage; people who wrote a pipeline may no longer work at the company or be so far removed from its operations that they'll need additional time to transfer knowledge (or re-discover it). You'll also start identifying rough timelines to execute the migration and potential change windows if you operate with code freezes or scheduled upgrade windows.

At the end of this stage, you should have a well-defined and prioritized backlog of work to be completed as part of the migration. While this may feel like over-planning or heavily process-oriented, it's a critical step, allowing you to ensure that all stakeholders are aware of the requiredmigration:sequence" activities, allowing them to prepare for the work to be done and any impacts on their day-to-day activities.

Migrate

Now that you've sorted all your planning, you can start doing "the work." This is a fairly straightforward set of activities that should be familiar to any developer. While you're doing this work, there are some guidelines we strongly recommend you pay attention to.

Minimum migration parity

Do the bare minimum to create workflows that have functional parity with your source workflows. Do not refactor patterns you currently have in place that work if you can create them in a one-to-one fashion in Airflow, even if it makes the code anti-pattern. *Ugly code that matches your source is better than new Airflow code.*

Identify tech debt

As you're working, you will identify tech debt that needs to be paid off. Do not attempt to pay it off mid-migration! Start tracking your tech debt items in a ledger, identifying what the debt is and how you can pay it off so that you can make decisions about when to do so at a later date.

Test heavily and completely

Reuse as much of your QA/testing cases as you possibly can. If you don't have any, now is a great time to document some based-on conversations with technical and business stakeholders. Test your workflows in isolated environments, using production data and scales if possible, and increase your testing if you work on a high-risk workflow where interruptions could be particularly disastrous.

Monitor

During the migration activity, you should be monitoring the status and progress of all activities. You should consider holding short daily meetings to escalate blockages or announcing hand-offs between teams. Ensure that all the stakeholders for the active migration set are invited and available to help clear blockers.

You should also maintain a **RAID (Risks, Actions, Issues, and Decisions)** log to track and communicate any issues and decisions that are made during the migration. This is a useful tool to ensure that "everyone is on the same page" about how the migration is progressing.

Use the daily status meetings and RAID log as the basis for any sort of weekly or monthly report to leaders and stakeholders. This also provides a convenient path for escalations should the need arise.

Technical approaches for migration

In this section, we will discuss a series of technical solutions to help alleviate commonly encountered activities as part of a migration. Not all of these solutions will apply to every scenario, but they're worth considering and deciding on as part of planning your migration.

Automating code migrations

If you have a large number of workflows to migrate, automating the generation of your Airflow DAGs from source artifacts will be a natural point of discussion and decision to be made.

When deciding to undertake automation, it is natural to try and make your tooling as perfect as possible in taking source code to a new destination – don't try for perfection. Remember that you're only going to do this migration "once," so you should just be shooting for "good enough" to get you most of the way, and accept that a certain number of manual steps will be required to clean up and finalize any code generation, creation of variables, and definitions of connections.

QA/testing design

If you have the luxury, take time to set up a formal QA, test plan, and environment for the migration. You should be interacting with two environments throughout this entire process – production and UAT in production.

Set up your UAT environment as a mirror of your production environment, but only provide read access to data sources. Workflows should read from production sources and write to isolated data stores in your UAT environment. As you're doing your testing program, you should compare your UAT data stores and performance to how your source workflows behave in production.

Once you have completed your testing and validation stages, promote workflows from UAT to production, and shut off the legacy workflow before activating your new Airflow DAGs.

Planning a migration between Airflow environments

As your Airflow instances grow, it is not uncommon to wish to migrate workloads to new environments. Sometimes, this is due to a change in service providers; other times, it is due to a need to share your workflows so that teams/workloads that once ran in the same deployment are isolated from each other.

All of our previous advice still holds true here, but because you're going from Airflow to Airflow, it's a generally much more straightforward activity and may not require as much testing to complete.

After identifying the DAGs you wish to migrate, you need to identify which objects you need to migrate to support their execution and whether you need to migrate previous execution history.

Connections and variables

Identify any connection or variable objects that you need for your DAG; if you're using a secrets backend, you will need to consult with that service for how to migrate variables to a new environment. If you're purely using some form of environment variable (or environment variable injection with some sort of Kubernetes secret), you should utilize your environment variable manager documentation to identify how to best migrate this data.

If you're using Airflow's metadata database to store these objects, you can utilize the core Airflow library to export (and potentially re-import) your objects of interest:

```
from airflow.models import Variable
from airflow.models import Connection
required_connections = []
required_variables = []
all_variables = session.query(Variables).all()
all_connections = session.query(Connection).all()
for v in all_variables:
```

```
if v.key in required_variables:
session.add(v)
session.commit()
for c in all_connections:
if c.conn_id in required_connections:
session.add(c)
session.commit()
```

Using the preceding pseudo-code, you can see how trivial it might be to export (and potentially re-import) data between two Airflow instances; you simply need to identify which variables and connections you need. The preceding basic pattern can be extended for additional objects that are represented as core data models within Airflow. If you have a small number of connections or variables to migrate directly between two Airflow instances, it's also reasonable to simply manually migrate them via the UI.

DAGs

Once you've migrated your core objects and shared your code to a new repository, you can determine how to best go about finishing your migration and cutting over workloads.

The simplest method is to turn off the DAG in your source environment, update your code to set the last run of your DAG as the new start date for the DAG, and then after pushing your code to your environment, activate your DAG in a new Airflow instance. This method is perfectly feasible if you have a small number of DAGs.

If your DAG doesn't have a start date, can't have its start date modified (and has catchup set to True), or simply can't lose its previous run history, you can migrate the information in the metadata database about previous DAG runs to the new instance, ensuring that the operational states of both instances are complete:

```
from airflow.models import DagRun
dag_runs = session.query(DagRun).filter(DagRun.dag_id = "dag_id")
```

To complete this operation, turn off the DAG in your source Airflow environment, utilize the preceding query to retrieve the current state of DagRuns executed for the DAG by ID, and insert it into the new environment before activating the DAG in the new environment.

Summary

Migrating workflows between tools and environments is an activity you are likely to find yourself doing at least a few times during your career as a data engineer. We hope that you now have the tools and techniques that you need to plan out and execute your migrations safely and promptly. We hope that no matter what tool you're migrating to or from, you find our basic guidance useful here. Remember that with activities such as this, it's better to spend the extra time planning before you execute than cleaning up something you broke.

Index

Packtpub.com

Subscribe to our online digital library for full access to over 7,000 books and videos, as well as industry leading tools to help you plan your personal development and advance your career. For more information, please visit our website.

Why subscribe?

- Spend less time learning and more time coding with practical eBooks and Videos from over 4,000 industry professionals

- Improve your learning with Skill Plans built especially for you

- Get a free eBook or video every month

- Fully searchable for easy access to vital information

- Copy and paste, print, and bookmark content

Did you know that Packt offers eBook versions of every book published, with PDF and ePub files available? You can upgrade to the eBook version at packtpub.com and as a print book customer, you are entitled to a discount on the eBook copy. Get in touch with us at customercare@packtpub.com for more details.

At www.packtpub.com, you can also read a collection of free technical articles, sign up for a range of free newsletters, and receive exclusive discounts and offers on Packt books and eBooks.

Other Books You May Enjoy

If you enjoyed this book, you may be interested in these other books by Packt:

Data Engineering Best Practices

Richard J. Schiller, David Larochelle

ISBN: 978-1-80324-498-3

- Architect scalable data solutions within a well-architected framework
- Implement agile software development processes tailored to your organization's needs
- Design cloud-based data pipelines for analytics, machine learning, and AI-ready data products
- Optimize data engineering capabilities to ensure performance and long-term business value
- Apply best practices for data security, privacy, and compliance
- Harness serverless computing and microservices to build resilient, scalable, and trustworthy data pipelines

In-Memory Analytics with Apache Arrow

Matthew Topol

ISBN: 978-1-80107-103-1

- Use Apache Arrow libraries to access data files both locally and in the cloud
- Understand the zero-copy elements of the Apache Arrow format
- Improve read performance by memory-mapping files with Apache Arrow
- Produce or consume Apache Arrow data efficiently using a C API
- Use the Apache Arrow Compute APIs to perform complex operations
- Create Arrow Flight servers and clients for transferring data quickly
- Build the Arrow libraries locally and contribute back to the community

Packt is searching for authors like you

If you're interested in becoming an author for Packt, please visit authors.packtpub.com and apply today. We have worked with thousands of developers and tech professionals, just like you, to help them share their insight with the global tech community. You can make a general application, apply for a specific hot topic that we are recruiting an author for, or submit your own idea.

Share your thoughts

Now you've finished *Apache Airflow Best Practices*, we'd love to hear your thoughts! Scan the QR code below to go straight to the Amazon review page for this book and share your feedback or leave a review on the site that you purchased it from.

https://packt.link/r/1-805-12375-0

Your review is important to us and the tech community and will help us make sure we're delivering excellent quality content.

Download a free PDF copy of this book

Thanks for purchasing this book!

Do you like to read on the go but are unable to carry your print books everywhere?

Is your eBook purchase not compatible with the device of your choice?

Don't worry, now with every Packt book you get a DRM-free PDF version of that book at no cost.

Read anywhere, any place, on any device. Search, copy, and paste code from your favorite technical books directly into your application.

The perks don't stop there, you can get exclusive access to discounts, newsletters, and great free content in your inbox daily

Follow these simple steps to get the benefits:

1. Scan the QR code or visit the link below

https://packt.link/free-ebook/978-1-80512-375-0

2. Submit your proof of purchase
3. That's it! We'll send your free PDF and other benefits to your email directly

www.ingramcontent.com/pod-product-compliance
Lightning Source LLC
LaVergne TN
LVHW081526050326
832903LV00025B/1644